ED
D0309369

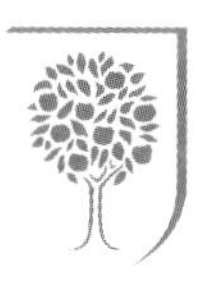

RHS

VEGETABLES
for the
GOURMET GARDENER

RHS Vegetables for the Gourmet Gardener
Author: Simon Akeroyd

First published in Great Britain in 2014 by Mitchell Beazley,
an imprint of Octopus Publishing Group Ltd,
Endeavour House, 189 Shaftesbury Avenue, London WC2H 8JY
www.octopusbooks.co.uk

A Hachette UK Company
www.hachette.co.uk

Published in association with the Royal Horticultural Society

ISBN: 978 1 84533 886 2

A CIP record of this book is available from the British Library

Set in Garamond Premier Pro, Frutiger and Edwardian Script

Printed and bound in China

Mitchell Beazley Publisher: Alison Starling
RHS Publisher: Rae Spencer Jones

Conceived, designed and produced by
Quid Publishing
Level 4, Sheridan House
Hove BN3 1DD
England

Designer: Lindsey Johns
Practical line illustration: Sarah Skeate

RHS consultant editor: Simon Maughan
Cookery consultant: Kate Heysmond
Food history consultant: Tracey Parker

The Royal Horticultural Society is the UK's leading gardening charity dedicated to advancing
horticulture and promoting good gardening. Its charitable work includes providing expert advice and information,
training the next generation of gardeners, creating hands-on opportunities for children to grow plants and
conducting research into plants, pests and environmental issues affecting gardeners.
For more information visit www.rhs.org.uk or call 0845 130 4646.

RHS VEGETABLES *for the* GOURMET GARDENER

Old, New, Common and Curious Vegetables to Grow and Eat

SIMON AKEROYD

MITCHELL BEAZLEY

CONTENTS

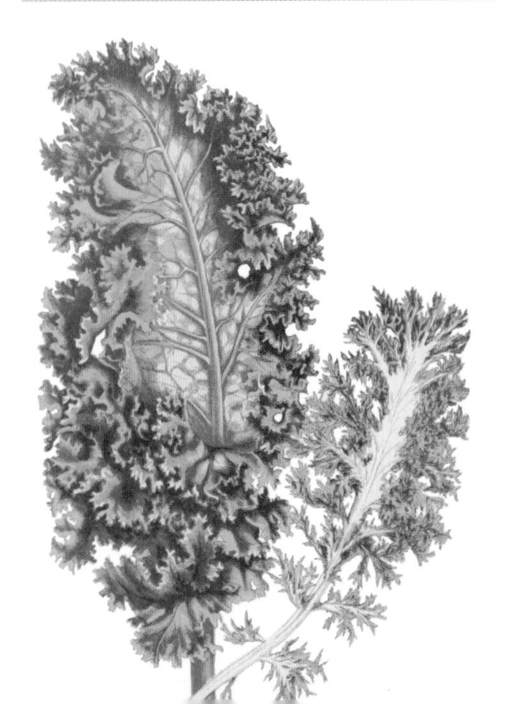

LEFT: One of the most common vegetables in Medieval Europe, curly kale (*Brassica oleracea* Acephala Group) dropped out of fashion until it was re-popularized during the 'Dig for Victory' campaign in World War Two.

ABOVE: The under-appreciated radish (*Raphanus sativus*) is ideal for kitchen gardens: quick to grow and easy to store, it is packed with vitamin C.

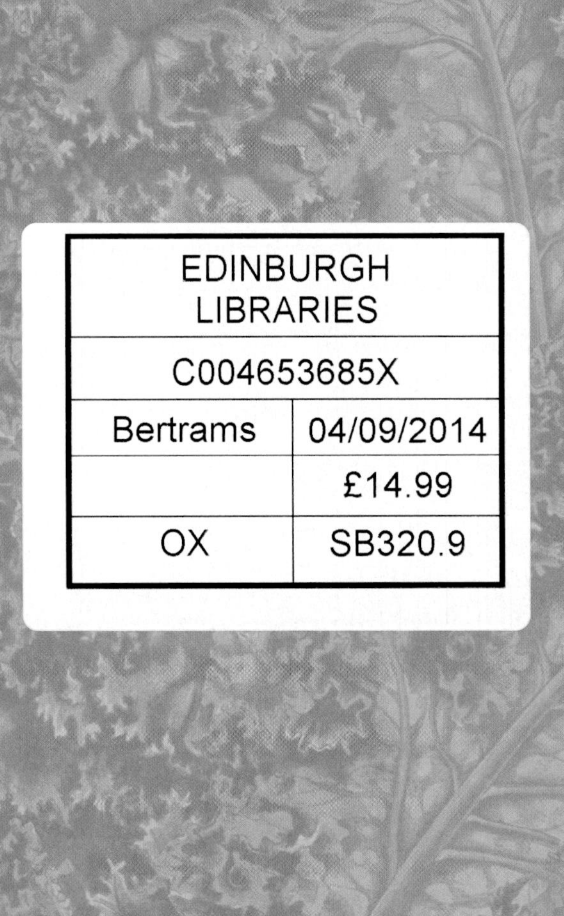

HOW TO USE THIS BOOK

THE VEGETABLES

At the top of the page in large letters are vegetables' common names, and in smaller italics the Latin name(s). Headings regarding plant type, climatic requirements, origins and brief cultural requirements are beneath in bold.

TASTING NOTES

Look out for the boxes with the knife-and-fork icon for culinary tips to identifying different varieties or cultivars for flavour, as well as simple gourmet recipes to try.

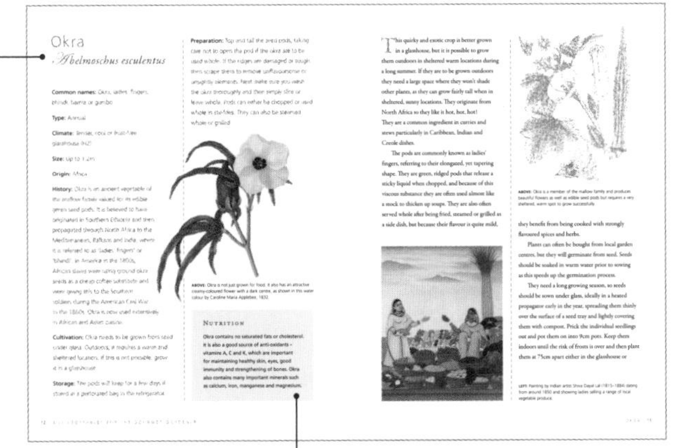

NUTRITIONAL INFORMATION

Health benefits specific to a vegetable are detailed in boxes.

PRACTICAL GUIDANCE

Line illustrations show hands-on explanations such as how best to garden or prepare vegetables.

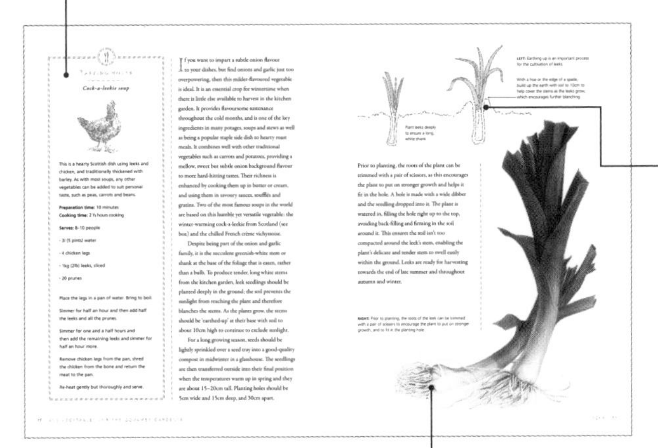

BOTANICAL ILLUSTRATIONS

There are more to vegetables than their green leaves. The beautiful illustrations in this book also highlight flowers, roots and seeds.

FEATURE SPREADS

The roles culinary vegetables can play in the garden or the kitchen are given additional detail in pages dedicated to traditional methods to put into practice.

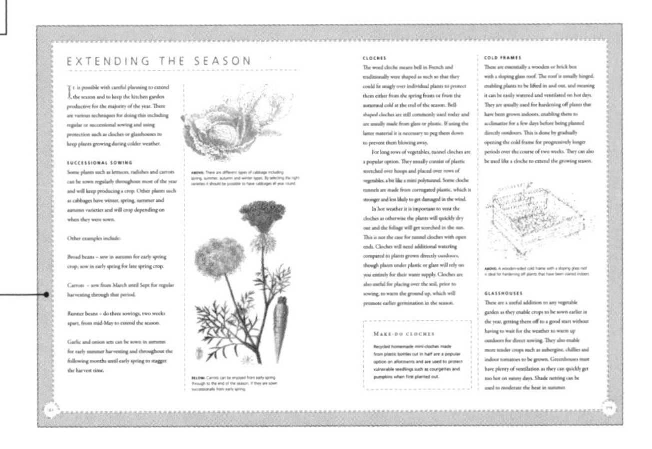

Introduction

Growing your own gourmet vegetables is guaranteed to take you on a life-absorbing adventure where the action occurs just a few metres from your back door. You will discover a range of new skills touching on geology, botany, horticulture and cookery, and you will learn to read the weather like a meteorologist and unearth incredible stories associated with historic and ancient varieties only found listed at the back of esoteric seed catalogues. Many such varieties have a long, exciting history that subsequently brought them to be common in our gardens and on our kitchen tables.

HISTORY OF THE HUMAN DIET

Mankind has toiled in the soil for thousands of years to produce its food. Growing vegetables was one of the first building blocks to creating ancient civilizations and societies, prior to which humans had been hunter gatherers, travelling around as they foraged for plants and animals. The ability to grow crops enabled them to settle down near fertile soil, cultivate land and build houses, villages and towns in the area. Following on from the building of houses, the next obvious transition was to create gardens where vegetables could be grown close to where they were to be cooked and consumed.

Many of the ancient techniques such as digging, sowing and weeding remain the same now as they did for our forefathers. In addition, the plants' requirements have not changed either – they still need the basic natural elements of sunlight, water and nutrients in the soil. But the one thing that has changed is the huge rise in popularity of kitchen gardening and allotments as people clamber to grow their own food.

Vegetables are packed full of healthy nutrients and goodness including potassium, folic acid and vitamins. Evidence shows they can reduce risks of heart disease, strokes, obesity, type 2 diabetes and cancer to name a few. Because vegetables are naturally low in fat and calories and they do not contain cholesterol, they provide nutritional food to improve people's health and well-being.

LEFT: 19th-century artwork for a series of adverts for a Parisian seed company, illustrating vegetables available at the time. The collection was finally published as *Album Vilmorin* (*Les Plantes Potageres*). This image dates from 1891.

With simply thousands of different vegetables to choose from, it should not be too hard to find even just a handful of vegetables that you enjoy eating regularly as part of a healthy, balanced diet for life.

ABOVE: 1879 artwork from the series of illustrations called *Album Vilmorin* (*Les Plantes Potageres*) from a Parisian seed company, Vilmorin-Andrieux & Co. The volume only survives in a few libraries today.

BECOMING A GOURMET GARDENER

As you gain experience, you will develop practical arts ranging from crop rotation basics to how to create a seed drill with the edge of your draw hoe. You will do battle with slugs, snails and tiny pests such as the carrot fly. Before too long you will start to treat your vegetable plot like a wine connoisseur treats his vineyard or cellar. You will find the subtle nuances of your plot, know which areas have the best soil or receive the most sun. You will recognize which crops to pick small and young to savour the best flavours, and which vegetables need time to mature like a fine wine. You will understand the best time to harvest and the optimum moment for storage to maximize the complex flavours.

In the kitchen, this book will show you how to transform these home-grown crops into delicious and sumptuous dishes. You will be able to hone your skills down to a fine art, and like a conjurer you will be able to magic up blue roast potatoes or purple carrots; impress yourself by growing lettuce leaves in winter and Brussels sprouts that do not taste bitter.

A gourmet gardener always has half their eye on the weather, with a brow to match the crooked furrow made by the rake in the soil. They know that all their hard work will always be in the hands of mother nature. Armed with fleeces and cloches in autumn and spring, and watering cans and shade netting in summer, the gourmet gardener becomes a master at adapting to the outside environment.

Whether you want the perfect recipe for making a cake mix or a compost mix, this book has it all. Once you start growing your own gourmet vegetables in the garden, it will take you on a learning journey that will become a rewarding hobby to last a lifetime. And this healthy exercise and eating can only contribute to a longer and more rewarding life. So what are you waiting for? Grab your wellington boots and start sowing.

WHY GROW YOUR OWN?

To the lover of gourmet vegetables nothing beats growing food from your own garden. Anyone who has grown their own food knows how much better it tastes than anything bought from a shop. Nothing rivals tasting a tomato warmed in the summer sun and picked fresh from the vine. And once tasted, who can forget the tender succulent flavour of asparagus harvested from the vegetable plot in the early morning dew and lightly steamed with butter for a breakfast treat? This is the experience that only a gourmet gardener can enjoy and embrace.

BELOW: Vegetables, such as these onions, come in many shapes and sizes. Growing your own enables you to cook with a range of flavours and colours not found in shops.

WIDER CHOICE

The vast range of vegetables available from seed companies cannot be matched on the shelves of the shops. Walk into a shop and there is a choice of about two varieties of onions. Open up a seed catalogue and there are often 15 or 20 varieties coming in all shapes, sizes and colours.

Suddenly there is a whole new world of exciting new vegetables to try which are almost impossible to buy in the shops. How often will you find blue potatoes or purple carrots, or be able to try the early spring hosta shoots as they unfurl from the ground or enjoy the subtle asparagus flavour from the asparagus pea? Only by growing these unique crops will a gourmet cook be able to embrace the full range of ingredients needed to make great food.

MAGICAL EXPERIENCE – FROM PLOT TO PLATE

Feeling part of the rhythms of life and embracing the seasons can only be felt outdoors. Feeling the soil in your hands and the sun on your back as seeds are sown and crops are harvested becomes an intoxicating experience. Growing food with the distinct flavour of the minerals and nutrients from your own soil gives the gourmet gardener magical

'In a world where we are becoming increasingly alienated from what we eat, growing our own vegetables is a fundamental way to reassert the connection between ourselves and our food'

Carol Klein, *RHS Grow Your Own Veg*, (2007)

ingredients that make the food produced completely unique to that location. Like alchemy, once in the kitchen those exclusive gourmet crops are transformed into great-tasting dishes that cannot be replicated anywhere else in the world.

HEALTHIER

For those with environmental concerns, there are of course no air miles involved with bringing your 'plot to plate'. As a gardener you have complete control over whether it is treated with chemicals or fertilizers or not. The physical exertion of growing gourmet vegetables is better than any gym work-out, and will make your muscles ache in a good way.

It is considered by many that home-grown veg has a higher nutritional value, far better for you than the produce of commercial farming practices that have squeezed the health and nutritional benefits out of the plants in a quest for uniformity and long-term storage benefits.

KEEPING OUR RICH HERITAGE ALIVE

If variety is the spice of life then growing your own gourmet crops is a must for anybody interested in growing and cooking food. Without that passion, all of those unique flavours, colours and varieties – many of which have wonderful historic stories attached to them – will be lost. The lover of gourmet food will be foraging back in the supermarkets with a choice of just a handful of uniform and often bland-tasting vegetables.

HISTORY FROM THE VEGETABLE PATCH

Some of the older and quirkier vegetables have unique flavours, colours, shapes and textures that are no longer in existence in the modern commercial vegetable world. By growing these gourmet crops you will help to keep them and the stories behind them alive. Often referred to as heritage or Heirloom varieties, the French call them by the evocative name *les legumes oubliés*, the 'forgotten vegetables'.

There are some wonderful stories attached to some of these historic vegetables. For instance, the French bean 'Cherokee Trail of Tears' commemorates the 1838 march of the displaced Native American Cherokee nation, who are said to have carried these seeds on the journey to their new homeland.

When Howard Carter excavated the tomb of the Egyptian boy king Tutankhamen in 1922, pea seeds were among the treasures he unearthed. Today the archeologically minded gardener can grow *Pisum sativum* 'Tutankhamen', which originates from the English Highclere Castle estate of Carter's patron Lord Carnarvon.

RIGHT: Growing your own food promises crops of exciting vegetables to grow all year round.

Okra

Abelmoschus esculentus

Common names: Okra, ladies' fingers, bhindi, bamia or gumbo

Type: Annual

Climate: Tender, cool or frost-free glasshouse

Size: Up to 1.2m

Origin: Africa

History: Okra is an ancient vegetable of the mallow family valued for its edible green seed pods. It is believed to have originated in Southern Ethiopia and then propagated through North Africa to the Mediterranean, Balkans and India, where it is referred to as 'ladies' fingers' or 'bhindi'. In America in the 1800s, African slaves were using ground okra seeds as a cheap coffee substitute and were giving this to the Southern soldiers during the American Civil War in the 1860s. Okra is now used extensively in African and Asian cuisine.

Cultivation: Okra needs to be grown from seed under glass. Outdoors, it requires a warm and sheltered location. If this is not possible, grow it in a glasshouse.

Storage: The pods will keep for a few days if stored in a perforated bag in the refrigerator.

Preparation: Top and tail the seed pods, taking care not to open the pod if the okra are to be used whole. If the ridges are damaged or tough, then scrape them to remove unflavoursome or unsightly elements. Next make sure you wash the okra thoroughly and then simply slice or leave whole. Pods can either be chopped or used whole in stir-fries. They can also be steamed whole or grilled.

ABOVE: Okra is not just grown for food. It also has an attractive creamy-coloured flower with a dark centre, as shown in this water colour by Caroline Maria Applebee, 1832.

Nutrition

Okra contains no saturated fats or cholesterol. It is also a good source of anti-oxidants – vitamins A, C and K, which are important for maintaining healthy skin, eyes, good immunity and strengthening of bones. Okra also contains many important minerals such as calcium, iron, manganese and magnesium.

This quirky and exotic crop is better grown in a glasshouse, but it is possible to grow them outdoors in sheltered warm locations during a long summer. If they are to be grown outdoors they need a large space where they will not shade other plants, as they can grow fairly tall when in sheltered, sunny locations. They originate from North Africa so they like it hot, hot, hot! They are a common ingredient in curries and stews particularly in Caribbean, Indian and Creole dishes.

The pods are commonly known as ladies' fingers, referring to their elongated, yet tapering shape. They are green, ridged pods that release a sticky liquid when chopped, and because of this viscous substance they are often used almost like a stock to thicken up soups. They are also often served whole after being fried, steamed or grilled as a side dish, but because their flavour is quite mild, they benefit from being cooked with strongly flavoured spices and herbs.

ABOVE: Okra is a member of the mallow family and produces beautiful flowers as well as edible seed pods but requires a very sheltered, warm spot to grow successfully.

Plants can often be bought from local garden centres, but they will germinate from seed. Seeds should be soaked in warm water prior to sowing as this speeds up the germination process.

They need a long growing season, so seeds should be sown under glass, ideally in a heated propagator early in the year, spreading them thinly over the surface of a seed tray and lightly covering them with compost. Prick the individual seedlings out and pot them on into 9cm pots. Keep them indoors until the risk of frosts is over and then plant them at 75cm apart either in the glasshouse

LEFT: Painting by Indian artist Shiva Dayal Lal (1815–1884) dating from around 1850 and showing ladies selling a range of local vegetable produce.

or outside in a very sunny, sheltered spot. If okra is to be grown outdoors, then it should be placed in a cold frame first for a few weeks prior to being planted in its final position. Tall plants will benefit from being staked as they start to grow. Tips can be pinched out to encourage a bushier plant.

Pods start to form in midsummer and should be harvested when they are about 10cm apart. If they are left too long they lose their viscosity and become stringy.

BELOW: The pods of okra are commonly known as ladies' fingers because of their distinctive, tapering length. Pods can be chopped or used whole.

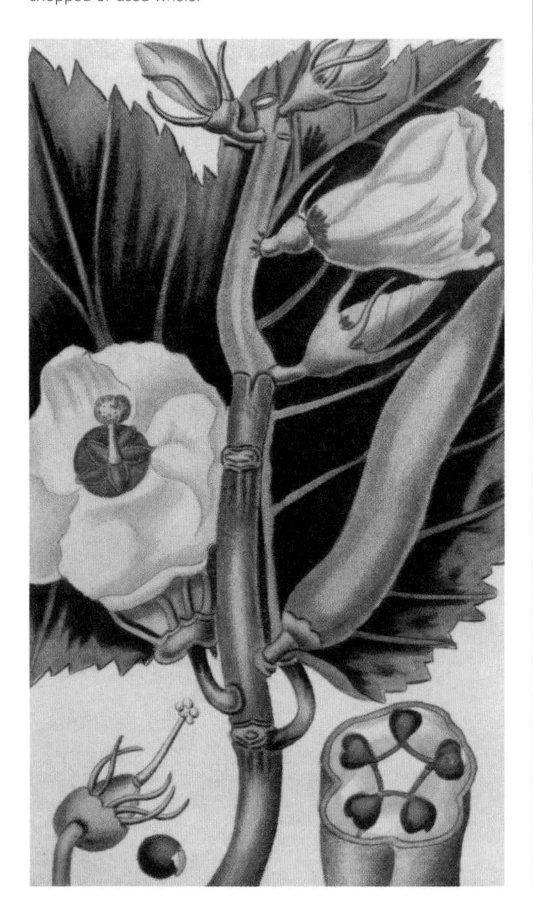

Tasting notes

Okra raita

Raita is a spiced yogurt dish with fried okra for added crunch and can be served as a side to accompany an Indian meal.

Preparation time: 15 minutes
Cooking time: 10 minutes
Serves: 2 people (as a side dish)

- 2 tbsp oil
- 250g (8oz) okra, washed and cut into chunks
- 1/2 tsp salt
- 1 fresh green chilli, seeded and chopped
- 125g (5oz) thick set natural yoghurt
- 1/2 tsp mustard powder
- 1/2 tsp black mustard seeds
- 1 tbsp curry leaves

Deep fry the okra in 1 tablespoon of oil in a frying pan until well browned and crisp.

Drain on kitchen towel and allow to cool.

Add the salt to the chilli and crush to a pulp.

Beat the yoghurt with a fork until smooth, add the mustard powder and chilli paste and mix well. Add in the fried okra.

Heat 1 tablespoon of oil in a small pan and fry the mustard seeds until they crackle, then add the curry leaves and fry for 15–20 seconds.

Remove the pan from the heat and stir the flavoured oil into the okra raita mix.

Elephant garlic

Allium ampeloprasum

Common names: Elephant garlic, Russian garlic, Levant garlic

Type: Bulb

Climate: Hardy, average to cold winter

Size: Up to 1.2m

Origin: Europe (particularly the Mediterranean area)

History: It originates from Central Asia where it has been grown for thousands of years, making it one of the most ancient vegetables ever cultivated. Archaeological and literary sources confirm its early usage by the Egyptians, Romans and Greeks as far back as 2100 BC.

Cultivation: Elephant garlic copes better with damper conditions than standard garlic, but still requires a well-drained soil in full sun. They are fully hardy and benefit from their individual cloves being planted in autumn as the cold winter promotes bulb development.

Storage: After harvest, leave them to dry in the sunshine for a week or two before storing. Trim the stem up to within 5–6cm. They should last for a few months if kept in frost-free, dry conditions. One handy tip is to store them in a nylon stocking and hang them up out of the way until needed in the kitchen.

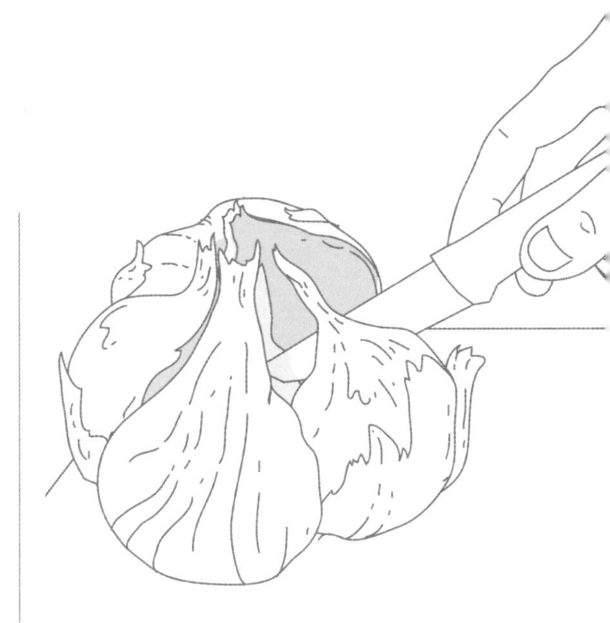

ABOVE: Elephant garlic bulbs are much larger than the usual garlic, and often a sharp knife is needed to prise the cloves apart.

Preparation: They can be cooked as a whole bulb or individual cloves. Due to its taste being milder than garlic, elephant garlic can be eaten raw in salads. Before eating in this way it is important firstly to peel away the papery skin, which can be removed easily. It is also possible to use cloves of elephant garlic as a vegetable – sliced and sautéed in butter or olive oil – as well as a flavouring agent. When cooking elephant garlic, be aware that it tends to brown even more quickly than other types of garlic, and this may give it a bitter taste. However, once cooked or grilled, elephant garlic takes on a gentle, sweet taste. It is lovely simply eaten on its own or spread on to a piece of crusty bread.

If you like cooking with garlic, but do not want to be overpowered by its pungent aroma, then give these closely-related but milder cousins a go.

Elephant garlic is so called due to the size of its enormous cloves. They are probably about four or five times bigger than standard garlic cloves, but what they make up for in size certainly is not reflected in the pungency and strength of their flavour. They are much milder than standard garlic and have a sweeter more nutty aroma, making them popular with cooks wishing to impart a subtle garlic flavour to their dishes. The bulb itself is the size of a large clenched fist, and will break up into individual cloves. Occasionally the bulb instead just forms one swollen bulb. These are called 'rounds' or 'solos' in regular gardener's parlance, and can still be chopped up and either cooked or used raw to flavour dishes. Alternatively, solos can be replanted the following autumn and will usually produce cloves.

Like other members of the onion family, it produces an attractive spherical flowerhead that can get as tall as 1.5m, making it an additional feature of a flower border, just as much as the vegetable beds. However, if you want large bulbs then the flower bud should be pinched out as soon as it starts to form.

Elephant garlic is planted exactly like standard garlic, pushing the cloves so that the tips are just below the surface. Due to their size, they benefit from being planted at a wider spacing, at 20cm between each clove in a row, and 30cm between the row. They tend to cope with slightly damper conditions, although the soil should still be well drained. They require a long growing season, so ideally they should be planted in autumn. However, they can also be planted out in early spring, but this can result in smaller cloves or producing 'rounds'. Keep the bulbs well watered as they start to grow and swell during spring. Harvest the swollen bulbs in summer when the foliage starts to turn brown and wilt, by carefully digging them up with a fork.

Tasting notes

Oven-baked elephant garlic

The less intense flavour and larger size of the elephant garlic lends itself perfectly to being baked. Delicious spread warm on crusty bread.

Preparation time: 10 minutes
Cooking time: 30 minutes
Serves: 2 people

- 1 elephant garlic bulb
- 2–3 tbsp olive oil
- Salt and pepper, to taste

Pre-heat a conventional oven to 200°C (400°F / gas mark 6 / fan 180°C).

Slice off the top of the garlic head and trim the bottom of the bulb so that it lies flat. Puncture with a fork.

Discard any loose skins and place the garlic bulb on a flat piece of tin foil.

Drizzle oil in the head of the garlic until it is filled. Season with salt and pepper and wrap tightly with the surrounding tin foil.

Place on a baking tray (or muffin tray) for 30 minutes. Baste several times during baking.

Once cool, peel the outside skins off of the bulbs and gently squeeze each clove out.

Leek
Allium porrum

Common names: Leek, poor man's asparagus

Type: Annual/biennial

Climate: Hardy, average to cold winter

Size: Up to 40cm

Origin: Mediterranean, Asia

History: Popular in ancient society, it was the Roman Emperor Nero's favourite vegetable as he believed it improved his singing voice.

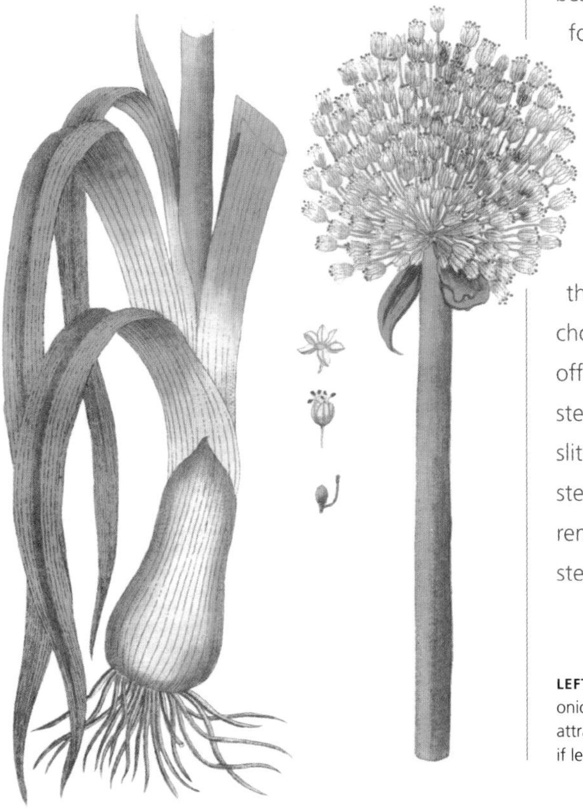

Leeks are one of the national emblems of Wales, possibly due to a Welsh legend of when King Cadwaladr of Gwynedd requested his soldiers should identify themselves in a battle against the Saxons by wearing leeks on their helmets. Welsh folklore has it that sleeping with a leek under the pillow will cause maidens to dream of their future husband.

Cultivation: Leeks prefer a well-drained soil, but reasonably rich and heavy. Add plenty of organic matter to light soil. Grow from seed in midwinter or buy seedlings in spring.

Storage: Leeks are best stored in nature's own larder – outside in the soil. They can be left growing in the ground throughout winter and best harvested before spring, as the centre forms a hard core rendering them inedible. They will store in the fridge for a couple of weeks. Freezing them will make them mushy, but they can still be used in soups and purées.

Preparation: Chop the untidy foliage off the top of the plant, and remove the roots by chopping off the very base of the stem. Strip off the outer layer to reveal the succulent white stem. Clean thoroughly to remove the mud by slitting the leek lengthways to the centre of the stem and rinsing with running water. Slice the remaining vegetable into sections and boil, steam or fry.

LEFT: Leeks are grown for their mild onion-flavoured stems, but they also produce attractive allium-like flowers (and seedheads) if left in the ground.

Tasting notes

Cock-a-leekie soup

This is a hearty Scottish dish using leeks and chicken, and traditionally thickened with barley. As with most soups, any other vegetables can be added to suit personal taste, such as peas, carrots and beans.

Preparation time: 10 minutes
Cooking time: 2 ½ hours
Serves: 8–10 people

- 3l (5 pints) water
- 4 chicken legs
- 1kg (2lb) leeks, sliced
- 20 prunes

Place the legs in a pan of water. Bring to boil.

Simmer for half an hour and then add half the leeks and all the prunes.

Simmer for one and a half hours and then add the remaining leeks and simmer for half an hour more.

Remove chicken legs from the pan, shred the chicken from the bone and return the meat to the pan.

Re-heat gently but thoroughly and serve.

If you want to impart a subtle onion flavour to your dishes, but find onions and garlic just too overpowering, then this milder-flavoured vegetable is ideal. It is an essential crop for wintertime when there is little else available to harvest in the kitchen garden. It provides flavoursome sustenance throughout the cold months, and is one of the key ingredients in many potages, soups and stews as well as being a popular staple side dish to hearty roast meals. It combines well with other traditional vegetables such as carrots and potatoes, providing a mellow, sweet but subtle onion background flavour to more hard-hitting tastes. Their richness is enhanced by cooking them up in butter or cream, and using them in savoury sauces, soufflés and gratins. Two of the most famous soups in the world are based on this humble yet versatile vegetable: the winter-warming cock-a-leekie from Scotland (see box) and the chilled French crème vichyssoise.

Despite being part of the onion and garlic family, it is the succulent greenish-white stem or shank at the base of the foliage that is eaten, rather than a bulb. To produce tender, long white stems from the kitchen garden, leek seedlings should be planted deeply in the ground; the soil prevents the sunlight from reaching the plant and therefore blanches the stems. As the plants grow, the stems should be 'earthed-up' at their base with soil to about 10cm high to continue to exclude sunlight.

For a long growing season, seeds should be lightly sprinkled over a seed tray into a good-quality compost in midwinter in a glasshouse. The seedlings are then transferred outside into their final position when the temperatures warm up in spring and they are about 15–20cm tall. Planting holes should be 5cm wide and 15cm deep, and 30cm apart.

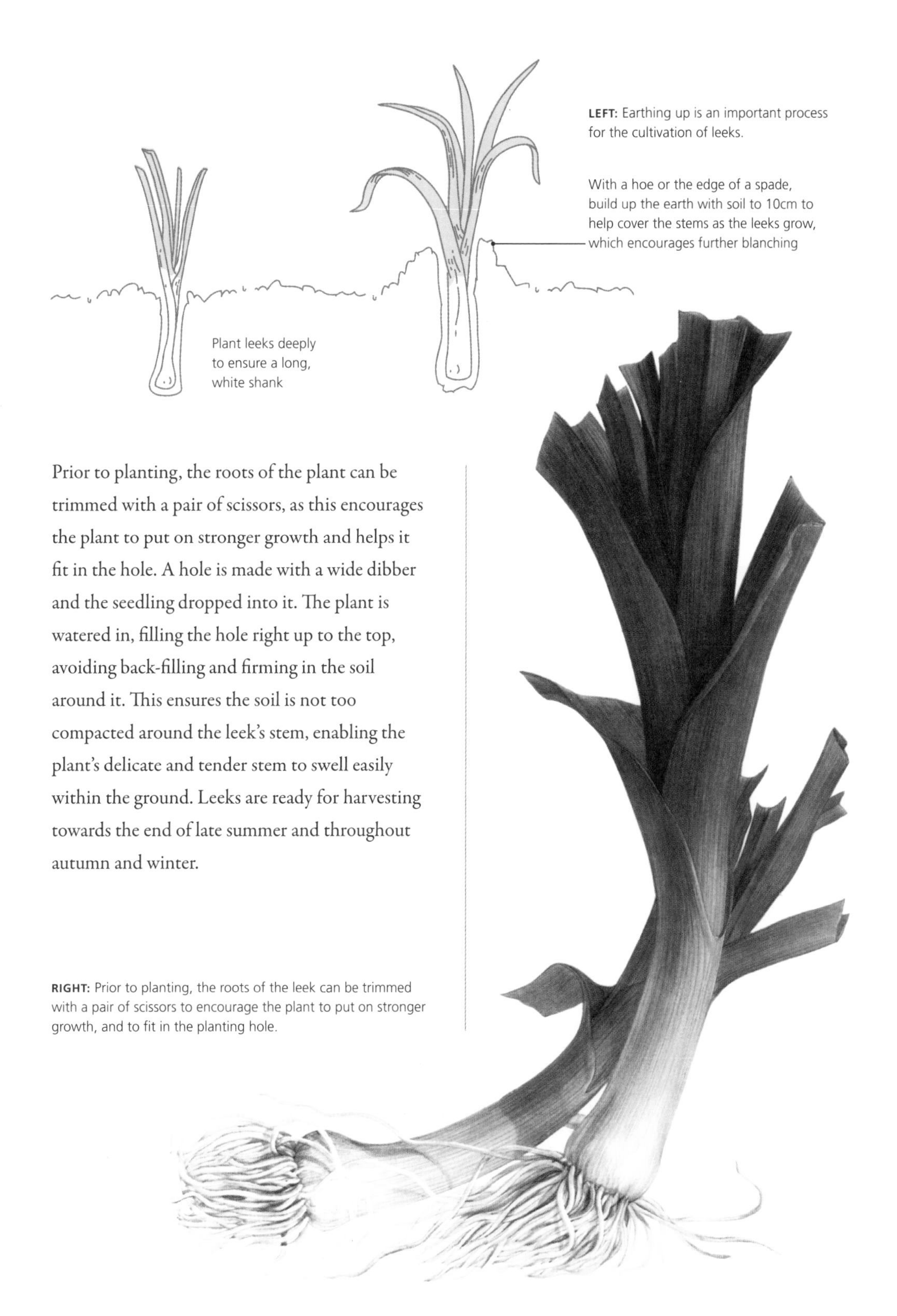

LEFT: Earthing up is an important process for the cultivation of leeks.

Prior to planting, the roots of the plant can be trimmed with a pair of scissors, as this encourages the plant to put on stronger growth and helps it fit in the hole. A hole is made with a wide dibber and the seedling dropped into it. The plant is watered in, filling the hole right up to the top, avoiding back-filling and firming in the soil around it. This ensures the soil is not too compacted around the leek's stem, enabling the plant's delicate and tender stem to swell easily within the ground. Leeks are ready for harvesting towards the end of late summer and throughout autumn and winter.

RIGHT: Prior to planting, the roots of the leek can be trimmed with a pair of scissors to encourage the plant to put on stronger growth, and to fit in the planting hole.

Onion
Allium cepa

Common names: Onion, bulb onion or common onion

Type: Bulb

Climate: Half-hardy, mild winter

Size: 40cm

Origin: Middle East

History: The onion is possibly one of the earliest crops grown, with records of its cultivation dating back some 5,000 years. Its anatomy of circles within circles led the Egyptians to believe it held powers of eternal life and it became a source of worship. Drawings and paintings of onions were found in Egyptian tombs and they buried their Pharaohs with onions too. Onions were also believed by many cultures to sustain life and prevent thirst. The onion is very hardy, can be grown almost anywhere and was therefore seen as an important food in early diets.

Cultivation: Onions should be grown in a sunny, well-drained soil. They are usually grown from sets, which are immature onions, but they can also be grown from seed.

Storage: Leave onions to dry out in the sun for a few days before bringing them inside. They can be platted together or placed in a nylon stocking and hung up in a cool dry place where they will keep for a few months.

Preparation: The best way to prepare an onion is first to cut a slice from the top and then peel off the skin. Cut the onion in half lengthways and chop each half separately.

To prevent eyes from watering when cutting onions, simply prepare them under running water or by placing them in a freezer for about 8–10 minutes before preparing.

Onions are indispensable to anyone who enjoys cooking. They are used worldwide to flavour a huge range of savoury dishes and can be eaten raw, but are more commonly fried. They can also be stored for years in jars of malt vinegar, and pickled onions are a popular healthy snack. Red onions tend to be milder and more popular in dishes requiring a raw onion, whereas brown and white onions are more often fried. Onions can be caramelized with sugar for a sweeter flavour.

LEFT: The onion is one of the most popular vegetables in the culinary world as it is used in such a wide range of dishes. It is also very easy to grow.

ABOVE: An image by French-Polish painter Jean-Pierre Norblin de La Gourdaine (1745 to 1830) showing a character with a traditional 'string' of onions around his neck.

Onions are usually grown from 'sets', which are simply immature onion bulbs. Once planted in the ground they start to swell, and after a few months of growing in the ground are harvested. They can be grown from seed too, but sets are easier and give the gardener a head start during the season.

Onions require a sunny site with well-drained soil. Avoid damp conditions as the bulbs are prone to rotting.

Onion sets should be planted out at 5–10cm apart in rows 30cm apart. Japanese and over-wintering onions can be planted in autumn for harvesting in midsummer, whereas onion sets should be planted in spring and harvested between midsummer and autumn.

Tasting notes

Onion bhajis

Onions are a common ingredient in Asian and particularly Indian cuisine. These easy-to-make onion bhajis are a great accompaniment to many of these dishes or they can be enjoyed on their own as a tasty snack or served on a bed of salad or pilau rice.

Preparation time: 20 minutes
Cooking time: 10 minutes
Serves: 4 people

- 2 eggs
- 3 onions sliced
- 120g (4oz) plain flour
- 1 tsp ground coriander
- 1 tsp cumin seeds
- 3 tbsp vegetable oil

Beat the eggs together, then mix in the onions.

Add the flour, coriander and cumin seeds and stir well. Heat the oil in a deep pan over a medium heat.

Add a large spoonful of the bhaji mixture to the oil and fry on each side for 30 seconds until it is golden brown.

Allow the bhaji to drain on kitchen paper.

Continue the process with the rest of the mixture.

Sowing seeds indoors

Onions can be sown from midwinter until mid-spring under glass. Although sowing seeds takes longer than growing from sets, it is cheaper and there is a wider choice of varieties to choose from. Modules should be filled with general-purpose compost. Sow a few seeds in each module at about 1cm deep and once they have germinated, they can be thinned out to the strongest seedling in each module. Plant them outside in spring.

Onions are ready for harvesting when the foliage starts to turn yellow and flop over. Gently lift out of the ground with a fork and leave to dry in the sun for a few days before taking them inside for storage.

Onions are relatively problem free, but there are a few pests and diseases to look out for. They can be prone to white rot, which produces a fluffy white appearance on the bottom of the bulb and causes the bulb to deteriorate rapidly. If this is seen, remove the bulbs immediately to prevent it spreading and avoid growing any other members of the onion family (garlic, shallots and spring onions) on the same ground for the next few years. Rotating your crops will help prevent this, and other infections in the soil. Growing carrots nearby should deter onion fly (see carrot fly on p.131), or covering them with a fine mesh should help prevent them. In damp, wet summers they can be prone to mildew. Regularly picking off the affected leaves should help contain the problem.

LEFT: This chromolithograph plate dating from 1876 by artist Ernst Benary shows a range of onion varieties taken from the *Album Benary*. It is one of 28 colour plates showing lots of different vegetables in cultivation at that time.

Spring onion

Allium cepa

Common name: Spring onion, scallion, salad onion, green onion

Type: Annual

Climate: Half-hardy, mild winter

Size: 25cm

Origin: Middle East

History: The word onion comes from the Latin word *Unio*, which means 'large pearl'. In Middle English, it became *unyon*. Eaten and cultivated since prehistoric times, onions were mentioned in the first dynasty of ancient Egypt as far back as 3200 BC. References to spring onions also occur in Chinese literature dating back over 2,000 years and they have also been used extensively in Chinese traditional medicines.

Cultivation: Like all members of the onion family, they require a sunny well-drained site. Seeds should be sown in spring.

Storage: Unlike onions and shallots, spring onions do not store for long. They will last for a few weeks if kept in the refrigerator or can be chopped and stored in the freezer.

Preparation: Cut off the roots and trim the green leaves to about 2.5cm (1in) above the white. Spring onions are almost always eaten raw and can be used whole or sliced in salads or chopped and put into stir-fried dishes.

ABOVE: Spring onions are a popular springtime vegetable and if left in the ground they will eventually produce clusters of eye-catching flowerheads.

These leafy, small bulb vegetables impart a mild onion flavour to savoury dishes. They are often referred to as scallions and are eaten or cooked fresh as they do not store for long. They are one of the first of the onion family to be harvested and as their name suggests they are picked from spring to early summer. They produce a thin stem of green foliage with a tiny white bulb at the base. Usually both the small bulb and the foliage are chopped up and added raw to flavour dishes, including salads and fish, and they taste particularly good in mashed potato.

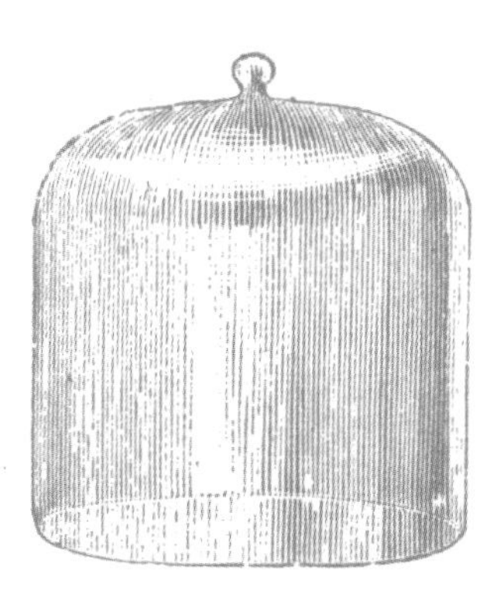

ABOVE: Placing spring onions under cloches in late winter means they can be harvested earlier in the season. Cloches are traditionally glass domed, but recycled plastic bottles work just as well.

Onions need to be grown in full sun in a well-drained soil. Manure should be added in autumn before sowing the seeds in early spring. Sow every three weeks throughout spring to early summer. They can also be sown in late winter and kept under a cloche for an early spring harvest.

Seeds should be sown thinly in shallow drills 15cm apart. They should not need thinning out as they are harvested when young and pencil thick. Avoid leaving them in the ground for too long as they tend to become tough quickly. They are usually ready for harvesting about 8 weeks after sowing. Remember to water them in dry weather and regularly remove weeds with a hoe to prevent the plants being deprived of nutrients and sunlight.

Spring onion bulbs are also occasionally used for pickling – although, for the genuine article, the cocktail or pearl onions should be used, which have slightly larger bulbs and a sweeter taste. There are a few varieties of spring onion worth trying. The most commonly grown variety is called 'White Lisbon', which is an old favourite with great flavour but prone to disease, particularly downy mildew. It is suitable for spring sowing and probably has the sweetest flavour of all the spring onion varieties. It is also one of the fastest growing. Another spring-sown variety worth trying is a hybrid called 'Laser' which is a non-bulbing type with white stems and good flavour.

Varieties suitable for sowing in late summer or early autumn, and that produce a crop the next spring, include 'Guardsman', which has some resistance to mildew, and 'Winter White Bunching' and 'Winter Over' which are hardier versions of 'White Lisbon'.

Tasting notes

Spicy cheese dip

This simple dip goes well with spicy sweet potato wedges. (See sweet potato on p.148 for recipe.)

Preparation time: 10 minutes
Serves: 10–15 people

- 1 spring onion, finely chopped
- 250g (9oz) cottage cheese
- Pinch of chilli flakes
- 1/2 tsp coriander

Simply mix all the ingredients together and spoon into a dipping bowl.

Shallot
Allium cepa var. *aggregatum*

Common name: Shallot, kanda, gandana or pyaaz (India)

Type: Bulb

Climate: Half-hardy, mild winter

Size: 40cm

Origin: Middle East

History: The shallot is a popular vegetable worldwide, particularly in French and Asian dishes. It is named after the ancient Palestine port of Ascalon, which is now modern Ashqueion.

Cultivation: Shallots grow in well-drained soil in full sun. They are usually planted as small bulbs or sets in early spring and harvested from mid to late summer.

Storage: Shallots will store for a few months after harvesting if kept in a dry, dark but cool place such as a cellar or garage.

Preparation: Shallots have a papery outer skin that should be removed first. Using a knife, cut off the root end and discard. They are often left whole in casseroles. Alternatively, they can be diced on a chopping board by making horizontal cuts into the shallot, almost to the root end. Then cut the shallot vertically into thin slices, holding it with fingers to keep its shape. Turn the shallot and cut it crosswise to the root end.

Shallots have just as rich a heritage as onions, have been cultivated for centuries and used worldwide in dishes for their intense and aromatic flavours. Less eye-watering in the kitchen than their bigger cousins, their taste can sometimes be more intense yet sweeter; they are often used to impart a more subtle background flavour to culinary dishes. They can be enjoyed raw in salads or added whole to stews and casseroles. They are also delicious when roasted. They come in a range of colours including yellow and brown, and they can be banana or torpedo shaped, but most of them are round. The pink ones tend to have the strongest flavour, and the red Thai shallots are popular in Asian cuisine.

RIGHT: Shallots are easy to grow from sets and are basically mini-onions, although rather than form one bulb, they form a cluster. They are grown for their aromatic flavours.

The main obvious difference between onions and shallots is their size. The other difference is that as the onion grows in the ground, it swells as one large bulb, whereas a shallot divides and multiplies into small clusters of bulbs.

Shallots are traditionally planted on Boxing Day by gardeners in the Northern hemisphere, but for many this may be too early and it is better to wait until late winter or early spring. Like onions, shallots are best grown from sets, which are immature or smaller onions. They should be spaced at between 10 to 15cm between each bulb in rows 30cm apart. Push them into the soil with the tip just showing above ground level. Keep them weed free as they grow and avoid watering them except in extremely dry periods, as the moisture can cause them to rot. Harvest shallots in summer by digging them up gently with a fork and leaving them to dry in the sun before bringing them inside to harvest.

A simple drying rack can be made to dry out shallots or onions immediately after harvesting. Stretch a piece of fine mesh between blocks of wood or bricks about 30cm off the ground. This will help the air circulate around the harvested crop. If available, place a recycled window pane or sheet of rigid plastic suspended above the bulbs to keep them dry in case of rain.

LEFT: Some shallot bulbs are unusual shapes and colours, such as this one with a red skin and mild sweet flavour.

Tasting notes

Red wine and shallot sauce

This sauce makes the perfect accompaniment to rump or sirloin steak.

Preparation time: 5 minutes
Cooking time: 30 minutes
Serves: 2 people

- 25g (1oz) butter
- 4 shallots, finely sliced
- 200ml (7fl oz) red wine
- 200ml (7fl oz) beef stock from 1/2 stock cube

Melt the butter in a large frying pan over a medium heat.

Add the shallots and cook for 2–3 minutes, to soften.

Add the red wine, increase the heat and boil for a few minutes to reduce by half.

Make up the beef stock and pour in.

Boil until reduced by half again.

Nutrition

Shallots have more anti-oxidants, minerals, and vitamins than onions. They are rich in vitamins A, B and E and essential minerals iron, calcium, potassium and phosphorus. Regular consumption of shallots can contribute to lowering cholesterol levels, improving the blood circulation and reducing the risk of cardiovascular disease.

Welsh onion
Allium fistulosum

Common names: Welsh onion, Japanese bunching onion, salad onion, perennial onion

Type: Perennial

Climate: Hardy, very cold winter

Size: 50cm

Origin: Asia; China

History: Despite its name, this onion is neither indigenous to Wales nor that popular in Welsh cuisine. The word 'welsh' hails from the old English word *welisc*, which simply means 'foreign'. The Welsh onion is believed to have originated in Asia.

Cultivation: Welsh onions can cope with partial shade or full sun. They can generally tolerate slightly moister soil than most onions as there is no bulb to rot.

Storage: They will store in the fridge for a couple of weeks. Alternatively they can be chopped up, placed into plastic bags or containers and frozen.

Preparation: Trim off both ends and chop in the same way as for a spring onion.

Probably the least known of the onion family, Welsh onions are a clump-forming perennial. The beauty of this plant is that it is fully hardy and can be harvested during winter and added to hearty warming soups and stews as a milder substitute to onions or shallots.

Welsh onions do not form a single large bulb but instead produce onion-flavoured stems with clumps of small bulbs beneath the ground, a bit like chives. They also produce an attractive flowerhead, making them a useful plant in the herbaceous border as well as the vegetable garden. Some varieties reach the size of leeks or larger while others more closely resemble chives.

Welsh onions are popular in Russia, where they are used in green salads. They are also commonly used in Chinese, Japanese and Korean dishes. It is the stem that is usually chopped up and eaten. They are usually eaten raw, but can be steamed or fried if they feel slightly tough.

LEFT: Welsh onions are a perennial type of allium, that can be harvested at any time of the year. It makes them a useful ingredient during the winter months.

Welsh onions are generally evergreen, but in particularly cold winters they will die back and regenerate new shoots in spring.

When it comes to harvesting, the stems or foliage can simply be cut from the plant as required. Alternatively, the entire plant can be dug up and bought inside for cooking. Due to their tendency to produce clumps of bulbs, even when the entire plant is harvested there are usually plenty of bulbs left in the ground to regenerate.

In spring, seed should be sown indoors into seed trays and kept on a sunny windowsill. In April they can be hardened off for a few days in a cold frame before being planted outdoors. Seedlings should be planted 10cm apart.

The other method of propagating them is by division. Simply use a spade to dig them out of the ground and pull apart their bulbs, replanting them in a fresh place in the vegetable patch. Bulbs should be planted so that their tips are just below the surface.

Walking onions

Walking onions, also known as Egyptian onions or tree onions (*Allium cepa* Proliferum Group), are a hardy perennial member of the onion family that produce small onion bulbils above ground in the tips of the plants. They have a milder flavour than onions, but are a useful addition to the kitchen garden as they will appear year after year once planted. They are called walking onions because the weight of the bulbils makes the plant flop over onto the ground, where they will start to root, and the whole process repeats itself.

Tasting notes

Cooking with Welsh onions

Welsh onions can be added to stews, casseroles and fish dishes after they are cooked. They must be used fresh and not cooked or their flavour will be lost. They are also good in salads and are lovely with butter in baked potatoes.

They can be used as an additional flavour in many other dishes to impart their mild onion flavour, substituting chives in an egg and mayonnaise baguette or a new potato salad, for example. Chop them up and add them to a sage and onion stuffing to accompany a Sunday roast chicken meal. Alternatively use them as a milder substitute to an onion bhaji.

Prior to planting Welsh onions, add plenty of organic matter, such as well-rotted horse manure or garden compost. This helps the soil to retain moisture, which will prevent the crop from drying out.

RIGHT: This is one of the lesser-known onions, which is a popular ingredient in Asian cooking. It has a variety of common names including Japanese bunching onion and Welsh onion.

COMPOSTING

The compost heap is the heart of the vegetable plot and the compost produced from it is often referred to by gardeners as 'black gold', and for good reason: every kitchen garden should have one.

A compost heap provides a green method of recycling all the waste produced from the kitchen garden. Once the waste has broken down, it provides material to help improve the soil. This can be dug into the ground prior to sowing or planting and can be used to mulch around fruit trees too, which will suppress weeds and retain moisture. Doing this reduces evaporation rates and is particularly useful in dry or poor soil. Compost adds a certain amount of nutrients to the soil and it also helps to aerate the soil as the organic matter attracts worms and other soil wildlife.

WHAT NOT TO ADD

- Avoid adding eggshells, meat and fish as this will quickly attract vermin and cause unpleasant odours that may affect you and your neighbours.
- Potato tubers will also quickly sprout in your compost pile. Boil them first if you want to add them to the heap.
- Perennial weeds will also quickly spread their pernicious roots through the heap. Instead, dry these plants out in the sun before adding fresh from the garden.

Recipe for perfect compost

Making decent compost is a bit like following a recipe. There are four essential ingredients: nitrogen, carbon, water and air. Ideally there will be about a 60:40 ratio between nitrogen and carbon.

Nitrogen	Herbaceous material such as grass clippings, stems, flowerheads and shredded foliage are the main sources of nitrogen. If there is too much green waste in a compost heap it will become smelly and slimy.
Carbon	Shredded paper, cardboard, autumn leaves, straw and shredded prunings from shrubs and trees provide carbon.
Water	If there is too much carbon in a mix, water may need to be added as this willl provide the extra nitrogen. Compost heaps should not dry out.
Air	Without air, the compost heap will not break down. Regularly turning the compost will speed up the process.

LEFT: Harvest leaves from the ground in autumn and winter and add them to the compost heap or make a devoted heap purely for leaf mould.

LOOKING AFTER THE COMPOST HEAP

- Turn the compost as regularly as you can – every 3 or 4 months if possible. The more you do it, the quicker the waste will rot down. Turning involves digging it out and placing it back into the heap, allowing the air to circulate around the heap. Having more than one heap is useful as the compost can be emptied from one into the other.
- If the weather is dry, the heap will benefit from being watered to speed up the decomposition.
- Regularly check on the compost heap and pull out any weeds or plants that start growing from it as they will spread.

TYPES OF COMPOST HEAPS

There are lots of different types of compost heaps that can be bought. These include drum types (dalek), rotating bins and sectional types.

A compost facility can simply be made by nailing or screwing three old pallets together to form a back and two sides. Pallets have gaps enabling air to circulate around the material.

Ideally there should be at least two compost heaps at different stages of decomposition. One that is ready to use, one in the process of decomposition. Three is optimum, as shown in the diagram below.

METHODS OF RECYCLING

There are other useful ways of recycling plant waste. These include worm bins and bokashi. Weedy plants, such as comfrey or stinging nettles, can be used to make a natural liquid plant food. To do this, simply steep the leaves in a bucket of water and leave them to rot for a few weeks. The resulting liquid can then be diluted and fed to plants.

Another technique is to make leaf mould from fallen deciduous leaves in autumn. Collect up all the fallen leaves and either place them in a punctured black sack or make a pile of them in a corner of the garden. They will eventually rot down, making a beautifully crumbly, rich black soil improver.

BELOW: Ideally there should be at least two compost sections on the go. One to keep adding material to, and one to be left to rot down. The system below shows an optional third section for the final rotting stage.

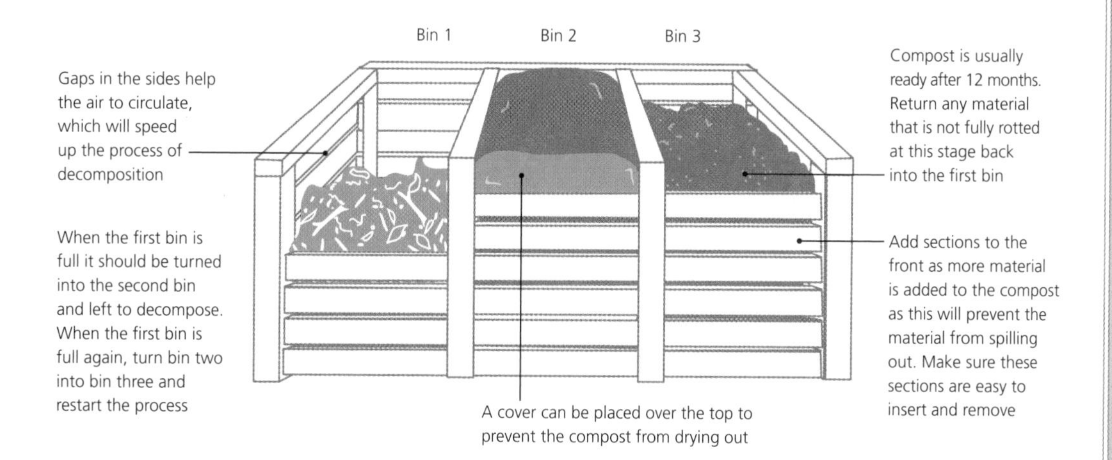

Garlic

Allium sativum

Common names: Garlic, garlick, rocambole

Type: Bulb

Climate: Hardy, average winter

Size: Up to 30cm

Origin: Central Asia

History: The word 'garlic' comes from the Anglo Saxon word *garleac* (*gar* meaning 'spear' and *leac* meaning 'leek'. It has a similar history to the leek and onion and can be dated back over 7,000 years to Central Asia. The ancient Egyptians worshipped garlic and placed models of garlic bulbs made out of clay in the tomb of Tutankhamen. The Romans believed that garlic held properties of strength and courage and fed it to their soldiers to give them the best start in battle.

Cultivation: Garlic requires a well-drained soil in full sun. It is fully hardy and its individual cloves are best planted in late autumn as the cold winter promotes bulb development, although it can be planted anytime through winter until early spring.

Storage: Leave bulbs out in the sun for a few days if the weather is dry, before collecting them up and storing them in a frost-free, dry place. They will store for 6 months or more.

Preparation: When choosing garlic it is important to look for bulbs that are hard and firm. The less papery the skin, the more moist the cloves will be. The papery skin should be removed and the cloves prised apart. The smooth skin surrounding individual cloves comes off more easily if gently crushed.

Garlic is one of the more pungent ingredients in the vegetable world, and just a tiny crushed clove is enough to flavour an entire dish. Anyone who has cooked with it will know how hard the smell can be to eradicate from the skin; it remains with the cook for hours afterwards. Popular in both Asian and Mediterranean cooking, garlic is a member of the onion family and is simply a bulb made up of usually between 8 and 12 individual cloves. The papery skin

LEFT AND ABOVE: Garlic is a popular bulb vegetable, closely related to onions, and with a pungent, aromatic flavour. It is used in a range of culinary dishes from around the world.

Nutrition

Garlic contains high levels of potassium, iron, calcium, magnesium, manganese, zinc and selenium, which are essential for optimum health. Garlic also contains health-promoting substances that that have proven benefits against coronary artery diseases, infections and cancers.

ABOVE: A coloured engraving of a peasant women c.1735 by Martin Engelbrecht illustrating a peasant lady with a variety of pink and white garlic bulbs attached around her waist.

that surrounds the bulb is usually white but there are attractive pink- and purple-tinged varieties too. Vampires might not be fond of this pungent bulb, but garlic is probably one of the most popular vegetables, with people from around the world using it to impart exciting flavours to otherwise bland dishes.

Garlic cloves are usually crushed or sliced in cooking, but they can be cooked whole. To provide a real punch of the garlic flavour, they can be added raw to salads, but breath mints will be required for hours afterwards if you do not want to upset your friends, family and work colleagues. For a milder garlic flavour, the stem or scapes can be harvested and cooked in stir-fries.

Tasting notes

Perfectly pickled garlic

Pickling reduces the powerful bite of garlic in its raw state, leaving a mellow and sweeter flavour. The result is great when used in sandwiches or with antipastas and salads.

Preparation time: 5 minutes
Cooking time: 10 minutes
Serves: makes a ½l (1lb) jar

- 48 garlic cloves, peeled
- 170ml (6fl oz) water
- 85ml (3fl oz) white or red vinegar
- 56g (2oz) sugar
- 1¼ tsp kosher salt
- ½ tsp whole black peppercorns
- ½ tsp mustard seeds
- ½ tsp fennel seeds
- ½ tsp red pepper, crushed

Bring a small saucepan of water to a boil over high heat. Add garlic and cook for 3 minutes.

Drain and place the garlic in a sterilized, heatproof, glass jar with a tight-fitting lid.

Combine the water, vinegar, sugar, salt, peppercorns, mustard seeds, fennel seeds and red pepper in a saucepan. Bring to a boil, stirring until the sugar and salt are dissolved.

Pour the hot pickling solution into the jar.

When cool, cover and refrigerate for at least 8 hours. Refrigerate for up to 1 month.

Garlic should be planted between late autumn and midwinter, ideally before Christmas. It is fully hardy and in fact requires a cold period of between 0–6ºC to encourage the bulb to develop.

Garlic requires a sheltered, sunny site with well drained soil. They struggle to grow on damp wet ground and will require plenty of grit or sand to be added if this is the case. Remove the papery covering from around the bulb and gently prise apart the individual cloves. These segments should then be pushed into the ground at 15cm apart ensuring that the basal plate (the flat section) is at the bottom. The tip of the clove should just be below the surface. Rows should be 30cm apart. Choose only the fat, plump bulbs for planting and discard any withered or thin ones. A net or fleece may need to be placed over them to if birds are attracted to the bulbs after planting. Avoid planting cloves bought from the supermarket as they may not be virus-free or suitable for the climate.

Hardnecks or softnecks

Garlic is divided up into two categories, hardnecks and softnecks, referring to the stem of the plant.

Hardnecks – the hardiest garlic, often producing a flower stem that can be cooked. They are usually planted out in autumn although they can also be planted in early spring. They generally have more complex flavours than the softnecks and a shorter shelf life, only lasting until midwinter time in storage. Varieties include 'Chesnok Wight' and 'Lautrec Wight'.

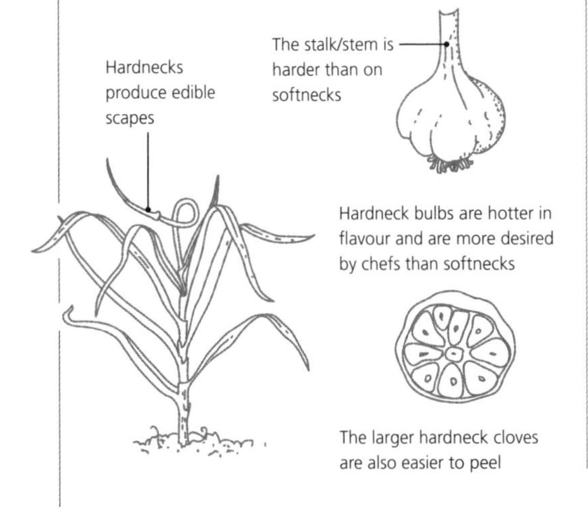

Softneck – this garlic contains more cloves, which are more tightly packed. They will last until late winter or early spring if stored correctly.

Varieties include 'Early Wight' and 'Solent Wight'.

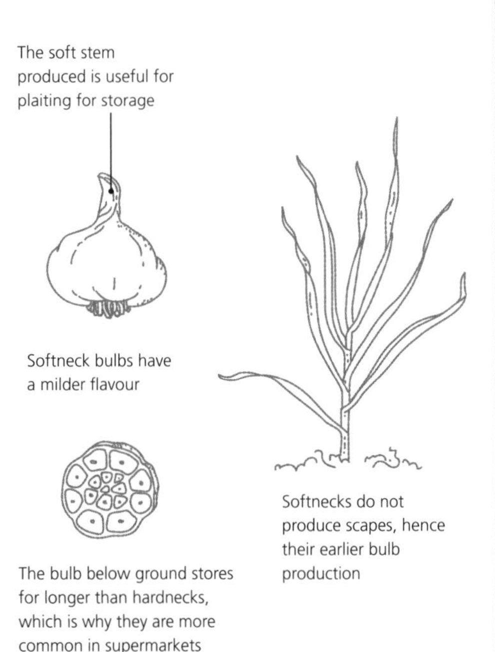

Planting garlic in modules

When the soil is heavy and wet, garlic can benefit from being planted out in modules in late autumn, left in a cold frame and planted out in early spring. Fill the modules up with multipurpose compost and push a clove into each individual cell, ensuring the tip is just below the surface. Place them in a cold frame, but ensure that the vents are open as garlic needs a cold period and keep them moist.

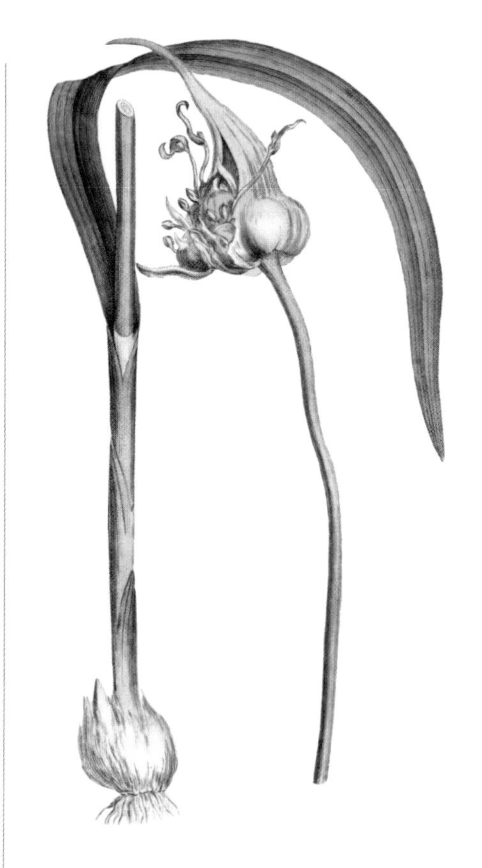

ABOVE: Garlic scapes are the 'flower stalks' of hardneck garlic plants that are often removed as they divert energy from bulb production, but can be added to dishes as they taste delicious.

After planting, water the bulbs during dry periods only – if overwatered, the bulbs may begin to rot. Regularly weed between the rows during the growing season as they are very susceptible to competition from weeds, and weak foliage is a sign of this. Remove any flower stems that may emerge from the bulb. Once the foliage turns yellow about midsummer, the bulbs are ready for harvesting. Unlike onion bulbs, garlic forms below the surface of the soil. Dig them up gently with a fork before the foliage dies down – otherwise it is next to impossible to know where they are in the soil – and leave them to dry in the sun for a few days.

Garlic can be stored in various ways for use during autumn and winter, including in net bags or by plaiting their stems together as you would a string of onions, leaving them to hang in a cool, dry place. The simplest and by far the most effective method is to thread a stiff wire through the base of the dry stem, adding one on top of another. They can then be hung up until needed. Avoid storing them in the kitchen as temperatures are often too warm.

> 'I must tell you that I have had a whole field of garlic planted for your benefit, so that when you come, we may be able to have plenty of your favourite dishes.'
>
> **Beatrice D'Este, letter to her sister Isabella, (1491)**

Love-lies-bleeding

Amaranthus caudatus

Common name: Love-lies-bleeding, love-lies-a-bleeding, pendant amaranth, velvet flower, foxtail, amaranth

Type: Annual

Climate: Tender to half-hardy

Size: 1m

Origin: South America

History: This plant was extremely popular in the decorative gardens of the Victorian era. Its genus name comes from the Greek word *amarantos*, which means 'unfading' with reference to its long-lasting blooms; its species name *caudatus* means 'with a tail'. This can clearly be seen from its wonderful long strings of cascading flowers. For many centuries amaranth was the principle grain crop of the Aztec people, who referred to it as 'the golden grain of the gods'. Its use was widespread on all continents. It has similar properties to corn and is classed as an ancient grain that was used long before wheat and corn became staple crops. In fact its seeds can be popped just like popcorn.

Cultivation: Requires a warm, sunny site outdoors but will tolerate fairly impoverished soil. It should be grown as an annual, sowing the seeds outdoors in spring, harvesting the leaves in summer, seeding in late summer and then adding the plant to the compost heap when the first autumn frosts arrive. Use the seeds in baking and other cooking.

Storage: The leaves do not store for long and can be kept fresh in the refrigerator for about 10 days. The seeds should be collected in paper bags and stored in a dry, cool and dark place such as a garage or cellar.

Preparation: Amaranth leaves can simply be washed in cold water and served as a salad vegetable. It can be cooked and served as a leafy vegetable similar to cabbage. Some gardeners prune larger plants for their tender leaves and tips. Others prefer to time plantings two weeks apart and pull up the young tender plants to eat.

LEFT: Love-lies-bleeding was a popular, decorative addition to the Victorian garden but is equally popular in cuisine. Its name is derived from its impressive cascade of bright red flowers.

If you want to brighten up your vegetable garden, then this is the 'must have' plant. Adding a touch of the exotic to the vegetable beds, it is commonly known as love-lies-bleeding as the flowerheads evoke a sense of blood being spilt from the plant. The attractiveness of its foliage and its brightly coloured catkin-like flowers that cascade down to the ground make it an architectural feature in the ornamental garden as well as the vegetable beds. It is sometimes grown as part of a bedding display. There are many different types of amaranth species, but the easiest one to find in seed catalogues is *Amaranthus caudatus*. It is very closely related to the grain quinoa.

Most of the plant is edible but it is the foliage and the seeds that it is mainly grown for. Leaves can be picked and either eaten raw or cooked like spinach. The seeds can be collected in paper bags and stored in a cool, dark but dry location like other grains until needed for cooking.

It requires a warm, sunny aspect in well-drained soil. The soil should be enriched with organic matter prior to planting or sowing although it will tolerate poorer soil than many vegetable plants. Seeds should be sown in spring under cover about four to five weeks before the last frost is predicted. Plant them out in late spring about 50cm apart. Alternatively they can be directly sown in the soil, scattering the tiny seed in the vegetable bed. As the seedlings emerge they can be spread to their final spacing of 50cm. The thinnings can be eaten like micro greens. Keep the plant well watered as they grow, and stake them with canes to prevent them flopping over. The seeds ripen towards the end of summer and can be harvested then. There can be as many as 100,000 seeds from one plant.

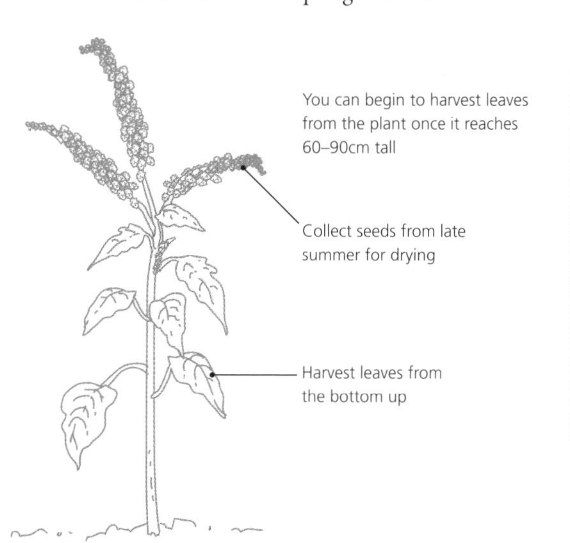

Tasting notes

Amaranth leaf pasta

Amaranth leaves are a healthy and flavour addition to a slightly spicy pasta dish.

Preparation time: 5 minutes
Cooking time: 15 minutes
Serves: 4 people

- 250g (1/2lb) whole wheat pasta shells
- 1 tbsp olive oil
- 150g (5oz) Amaranth leaves, chopped
- 3–4 cloves garlic, minced
- 1/2 tsp dried red pepper flakes
- Parmesan cheese, freshly grated, to taste

Cook the pasta, drain and put to one side.

Heat oil in a large pan over medium heat. Add the leaves, garlic and red pepper flakes; sautée for 5 minutes or until the garlic turns light gold. Add cooked pasta and mix well.

Celery

Apium graveolens var. *dulce*

Common name: Celery

Type: Annual

Climate: Hardy, mild to average winter

Size: 45cm

Origin: Mediterranean

History: The name celery is believed to be derived from the French word *celeri*, which in turn comes from the Greek version of the word. Celery is believed to have originated in the Mediterranean although other countries lay claim to this. Originally used for medicinal purposes as a flavouring herb, it was not until the 1600s in France that celery was seen to be actually edible and began to be used in cooking. The Greeks believed that celery was a holy plant and archaeologists found drawings of celery present in ancient Egyptian tombs.

Cultivation: Celery requires a moist soil but will tolerate some moderate shade. Seeds are sown in spring and the crop is usually harvested in autumn. Stems are traditionally blanched to sweeten them by covering them up with soil, but there are modern self-blanching varieties available.

Storage: Celery can be stored in a perforated bag in the fridge for around 10 days. If blanched for three minutes in boiling water, it can be frozen for later use, but it will have lost its crispness and is only suitable for cooking.

Preparation: Stems can be broken off when ready to use. The outer stems are tougher and are better for cooking, whereas the inner ones are more tender and suitable for eating raw. The leafy tops can also be used in salads. Wash thoroughly to remove the soil from the base of the stems.

LEFT: Celery was originally used as a flavouring herb to make medicines more palatable, but it is now an important addition to the cook's list of ingredients.

Celery has been putting the crunch into salad for the last few centuries. It consists of a cluster of long juicy stems grown around a central heart. Stems vary from green to white, the lighter stems being sweeter and less bitter. The flavour is unmistakable and reminiscent of very mild onions. The stems are crisp and popular with dieters as it is claimed more calories are burned eating this succulent vegetable than are consumed. The best way to eat celery is simply to chop off a stem and eat it whole, some people preferring to shake salt over it. However, stems can be sliced diagonally and mixed into stir-fries, or chopped into 1cm lengths and added to soups and stews. The leafy tops are often used to add flavouring to dishes.

Growing celery has fallen out of favour in recent years as it has a reputation for being tricky to grow. It is a vegetable that requires constant moisture to thrive and can be considered hard work to dig out of the garden trenches. It is a shame, as home-grown celery is far superior to most of the bland varieties for sale in the shops. There are green, red and pink-stemmed varieties available.

Celery always tastes better when the sunlight has been excluded from the stem. This is called blanching and there are two methods of doing this.

To blanch celery the traditional way, a trench should be dug 45cm wide and 30cm deep in autumn. Add well-rotted manure to the bottom of the trench so the trench is left at 10cm deep. Seeds should be sown under glass in spring in a heated propagator. Once they have been hardened off in a cold frame for a few days, they should be planted in the trench at a spacing of 30cm apart. Rows should be at 60cm apart. As they start to grow, the stems should be loosely tied together just below the leaves, and the soil earthed up around the stem with a draw hoe. This should be repeated every few weeks during the growing season until the soil has reached up to the lower leaves.

The other method of blanching is less labour intensive. Rather than digging out a trench, the celery is simply planted out at 30cm apart, and cardboard collars or waterproof papers are wrapped around the stems every three or so weeks, as the stems grow.

Nowadays, there are self-blanching celery varieties available, although some claim that they do not taste as good. They should be planted in blocks rather than rows as their dense foliage prevents the sunlight from reaching the neighbouring plants, which is how the stems are blanched.

It is essential that celery is kept well watered throughout the growing season. Regularly weed between the rows to prevent them competing for moisture with the plants.

Celery is ready for harvesting in autumn. The plant can remain in their trench until ready to use in the kitchen, but should all be harvested before the weather becomes too cold.

RIGHT: Celery is easier to grow than you think, and modern varieties do not need blanching to create white stems.

Celeriac
Apium graveolens var. *rapaceum*

Common name: Celeriac, knob celery, celery root, turnip-rooted celery

Type: Annual

Climate: Hardy, average winter

Size: 35cm

Origin: Mediterranean

History: Celeriac is derived from wild celery, which has a small, edible root and has been cultivated as an edible plant for thousands of years. In the Middle Ages, it spread from the Mediterranean, finding its way into Northern European cuisine.

Cultivation: Liking a sheltered position in sun, celeriac will tolerate moderate shade. It requires a long growing season to fully mature and, like celery, needs a moist soil to thrive.

Storage: The best way to store celeriac is in the ground until it is needed. It can be stored in a cool dark place for a few weeks, but once it is chopped up it needs to be used as it discolours rapidly.

Preparation: Cut off the top and base of the celeriac then remove the coarse, tough skin with a knife or sharp potato peeler. Sections of it can then be chopped as required.

LEFT: This ugly-looking vegetable, to the right of the celery, is often called turnip-rooted celery as the base of the plant looks like a turnip, yet the flavour of the swollen stem is reminiscent of celery.

This knobbly vegetable would not win a beauty contest, but what it lacks in looks it makes up for with its versatility in the kitchen. It is not just the name that is similar to celery; the flavour is reminiscent too with slightly nutty overtones. It makes a great celery-flavoured substitute for winter dishes, when fresh celery will not be available in the vegetable garden, and it can be grated and fried up to add to salads. Alternatively, it can be treated like a potato and chopped into large chunks and boiled for 20 minutes or baked or roasted in the oven for 40 minutes. Celeriac mash with garlic and cream is a delicious winter-warming treat.

LEFT: It is possible to harvest the swollen stems when young, once they reach anything from 10cm across.

RIGHT: 'Giant Prague' celeriac is a heritage variety first grown about 1871, grown for its large, white roots.

Celeriac is much easier to grow than celery but it does need a long growing season in order to develop its swollen stem. They are quite hungry plants so the soil should be enriched with plenty of organic matter and left to settle for several weeks prior to planting. Seeds should be sown under glass in modules or small plastic pots in late winter or in early spring. Place them in a propagator at 15°C to encourage germination. When the plants reach 10cm tall they can be hardened off in a cold frame and planted out at 30cm apart in rows 30cm apart. The base of the stem should be level with the surface of the soil.

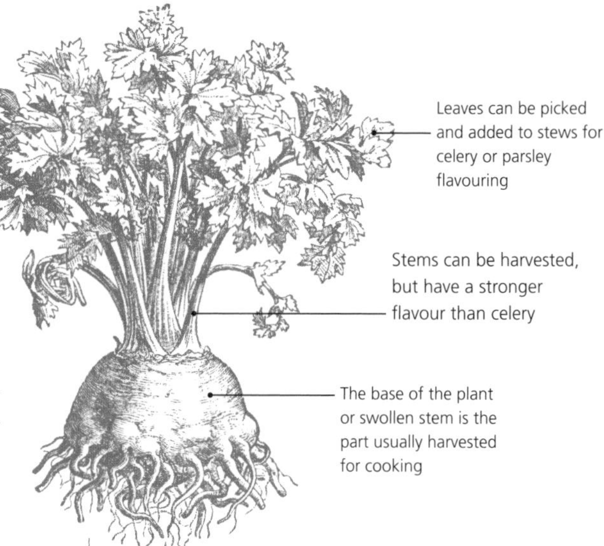

As the plant grows and the stem swells it is important that the soil is kept moist during dry conditions. Remove side shoots as soon as they appear as these will compete for nutrients. Also remove some of the lower leaves in late summer and earth up around the stem. In cold areas, as autumn approaches the plants will benefit from being mulched around their base with straw.

Harvest the swollen stems during autumn and winter as required using a fork to gently lift them out of the ground. Celeriac should be left in the ground for as long as possible to allow the stem to swell. However, they can be harvested and eaten when they are about 10cm across.

In cold areas where there is a risk of the ground freezing over and becoming difficult to dig, the celeriac can be lifted and stored in trays covered in moist sand and kept in a dark, frost-free place such as a cool garage or cellar. Regularly check the crops over and throw out any that spoil.

Tasting notes

Celeriac remoulade

The texture of celeriac is ideal for making a simple remoulade, which is a classic French dish of grated celeriac in a mustard-flavoured mayonnaise. It is great for accompanying fish dishes or can be served with toast.

Preparation time: 10 minutes, plus 2 hours for chilling
Serves: 4 people

- 1/2 small celeriac, cut into thin strips
- 2 tbsp double cream
- 3 tbsp mayonnaise
- 2 tsp Dijon mustard
- Lemon juice, to taste
- Salt and pepper, to taste

Mix the ingredients together and place in the fridge for a couple of hours before serving.

Asparagus

Asparagus officinalis

Common names: Asparagus, speargrass, spargel

Type: Perennial

Climate: Hardy, average to cold winter

Size: Up to 1.5m

Origin: Europe and particularly the Eastern Mediterranean area

History: There are references of asparagus dating back to the ancient Egyptians, Greeks and Romans. Popular throughout history, not just as a vegetable but also for its medicinal properties and use as a diuretic. Probably due to its shape and high phosphorous content, it has also had a reputation as an aphrodisiac throughout history.

Cultivation: Asparagus plants require a well-drained soil. On heavy soils with poor drainage, they will benefit from being grown in raised beds where excess moisture will drain away.

Storage: Stand them on their base in a glass of water and keep in the fridge for a few days. Trim and briefly blanch them with boiling water prior to freezing, either as whole spears or chopped.

Preparation: Remove the base of the stem with a knife, or simply snap it off. Older thicker stems can be peeled. Asparagus is usually briefly steamed or boiled for a few minutes. It can also be roasted or eaten raw by using a vegetable peeler to strip off thin lengths from spears and added to salads. It can be grilled over hot charcoal and often stir-fried in Asian cuisine.

LEFT: Asparagus is an early spring time treat. The emerging shoots are harvested when they are between 12 and 20cm tall. Shoots can vary in thickness depending on growing conditions and variety.

Nutrition

Asparagus contains asparanin, high levels of vitamins C and E, together with the minerals zinc, manganese and selenium, which all help to provide anti-inflammatory and anti-oxidant health benefits.

The season is short but the memory of these gourmet treats will linger for much longer. Often considered a luxury crop and backed up by luxury prices in quality food stores, growing them is surprisingly easy and will reward a gardener with succulent, tender spears for many years after planting.

The emerging spring shoots of this herbaceous perennial are often referred to as spears. In fact the word *asparag* is derived from an ancient Persian word meaning 'spear'. They are usually green but occasionally white when the emerging shoots have been covered to exclude light. It is also possible to grow purple varieties.

The spears have a refreshingly delicate yet herbaceous flavour and epitomize spring dishes. They are the perfect accompaniment to pork, chicken and fish, particularly salmon. To maximize their fresh, spring flavours, simply steam them for a couple of minutes, and add melted butter, a splash of lemon juice and season with cracked black pepper. The skill in steaming asparagus is all in the timing. Too quickly and the spears are al dente and too hard. Too long and the spears turn to mush. Yet the difference between the two extremes can be a matter of seconds in the steamer.

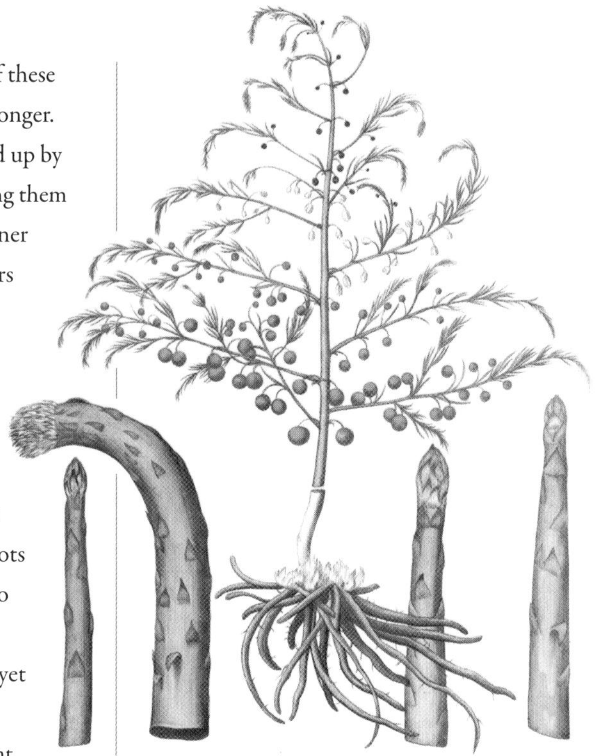

ABOVE: Harvest shoots regularly as they will quickly turn thick and unpalatable. Asparagus also produces attractive ferny foliage and berries in late summer and early autumn.

The ferny foliage and berries make an attractive late-summer and autumnal feature to the kitchen garden, but may require staking with strings to prevent them flopping over. Give careful consideration to where you want to grow these plants, as they are a long-term crop and could remain in the same bed for 15–20 years.

Patience is a virtue and this is certainly the case with this delicious spring delicacy. You should wait two or three years before harvesting the first

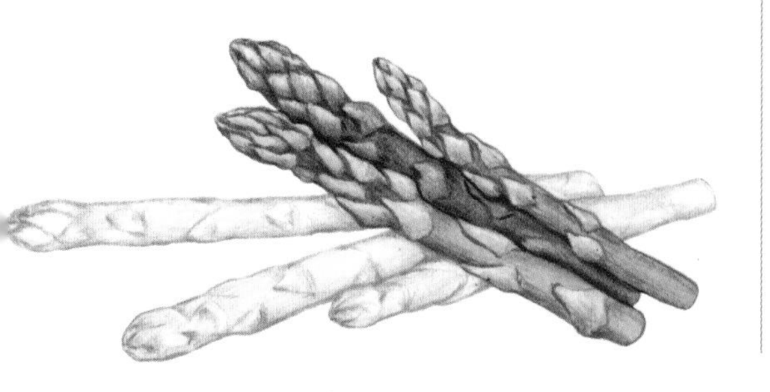

LEFT: Asparagus spears are more commonly green, but the stems can be covered with soil to blanch them and cause them to go white.

delicate spears after planting, which will allow the plants to get properly established.

Harvest the spears when they are between 12cm and 20cm tall. They grow quickly, so the beds will need checking for ready spears every two or three days to avoid any wastage. Cut the spears just below the surface of the ground using either an old knife or a special asparagus knife. Ignore any thin spears that appear and let them grow.

Asparagus is usually planted as small one-year-old crowns in early spring. Dig out a trench 30cm wide and 20cm deep. Add lots of well-rotted manure into the base and then create a 10cm high ridge along the trench. The crowns should be planted along the ridge 30cm apart and the roots should be spread downwards on either side. Partially backfill the trench, gradually refill as the spears begin to emerge and then completely refill when the spears are above the top of the trench. Rows should be 45cm apart and the crowns should be staggered with the rows on either side. Cut back the foliage to near-ground level in autumn when it starts to turn yellow and die back.

BELOW: The common method of planting asparagus is to create a mound in a trench and place the plant on top of it, spreading the roots around it.

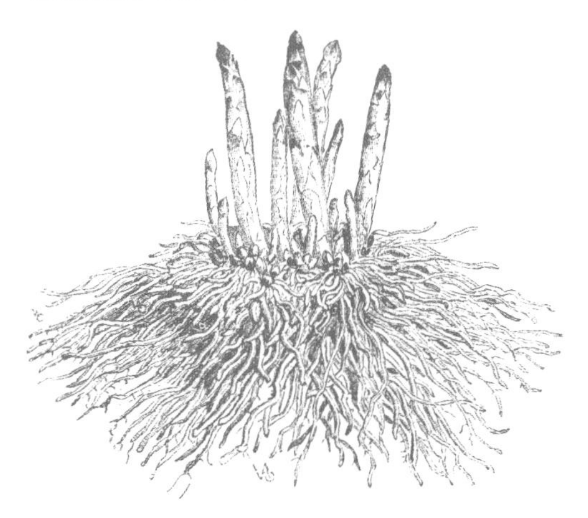

Tasting notes

Asparagus soup

A simple way to capture the delicate flavours of asparagus is to create a smooth creamy soup from freshly harvested spears.

Preparation time: 10 minutes
Cooking time: 20 minutes
Serves: 4–6 people

- 1 tbsp olive oil
- 1 onion, finely chopped
- 400g (13oz) fresh asparagus spears
- Knob of butter
- 500ml (16fl oz) vegetable stock
- Salt and pepper, to taste
- 4 tbsp double cream

Heat up the oil in a pan. Add the onion and fry for about 5 minutes until soft.

Add the asparagus spears with the butter, cover and sautée for 10 minutes.

Make and pour in the vegetable stock. Bring to the boil. Season and then simmer for 5 minutes.

Drizzle in the double cream and use a hand blender to mix together to a smooth soup.

SOWING TECHNIQUES

Most vegetables can be bought as plants from local garden centres. However, seeds are much cheaper to buy and there is a far greater range of varieties available. Furthermore, it is far more satisfying to grow something from seed, nurture it and finally harvest the vegetables. With many plants you can also harvest the seed and use it next year.

There are no hard and fast rules for seed sowing. At the end of the day, seeds simply need soil, water and sunlight. However, learning a few basic seed-sowing techniques should ensure the vegetables get off to the best possible start.

DIRECT SOWING

Many seeds can be sown directly into the soil. The packets of seed will provide information on the distance between each plant, between each row and the depth it should be planted at. Some seeds, such as broad beans and parsnips, are hardy and can be planted out in colder weather; other plants, such as runner beans, should not be sown until the soil has warmed up in early summer.

DIRECT SOWING OF SMALLER SEEDS

Some of the smaller seeds such as carrots. radishes, parsnips and lettuce require a shallow drill to be made into which the seeds are evenly sprinkled. Sand can be mixed with small seeds to make them simpler to spread. Once these seeds have germinated they will need to be thinned out to their final spacing as recommended on the seed packet. If they are not thinned out, then the vegetables will remain small, which some chefs prefer.

DIRECT SOWING OF LARGER SEEDS

Larger seeds, such as beans and peas are often individually placed into the soil using a dibber. After the seed is dropped into the hole, it is backfilled with soil. Occasionally larger seeds are planted in clusters and once germination has occurred the weaker ones are removed, allowing the stronger one to grow away.

LEFT: Parsnips should be sown directly into the ground in light soil, as their long tap root makes them difficult to transplant from pots.

'We plough the fields, and scatter the good seed on the land
But it is fed and watered by God's almighty hand.'

Matthias Claudius, *Garland of Songs*, (1782, published 1861)

SOWING INDOORS

Some seeds, such as tomatoes and chillies, need to be sown indoors as they need a long season to grow, yet need warmth to provide germination. Usually one or two seeds are sown into individual plastic pots or in modules. They should be left to germinate in a sunny, dry place such as a window ledge or glasshouse. After germination, the seeds can usually be thinned out to the strongest seedling. Large seeds such as pumpkins, squashes and courgettes are often sown on their side to avoid the water sitting on the surface, causing the seed to rot.

HARDENING OFF

Seeds that have been grown indoors benefit from being hardened off before being planted outdoors. This involves placing the seedlings in the porch or cold frame to toughen them up, preventing them from going into shock when being planted outside.

SOWING INTO SEED TRAYS

Seeds such as leeks are often sown indoors in seed trays full of compost. Seeds are lightly sprinkled over the surface, before being lightly covered over with more compost and watered.

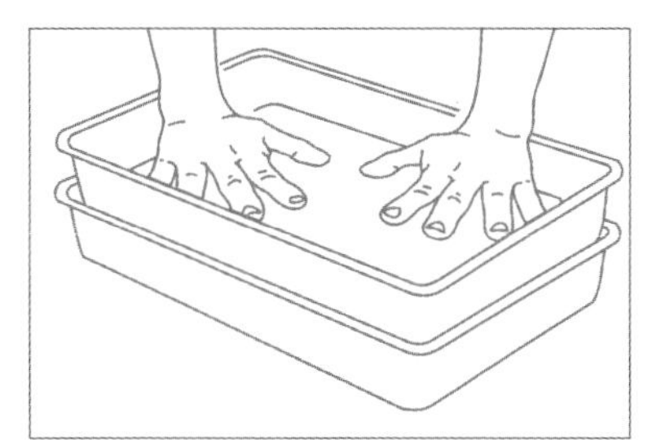

STEP 1: Fill a container with a good quality sowing compost until it is just below the top of the rim. Gently firm it down using finger tips, or a piece of wood, or the base of another tray.

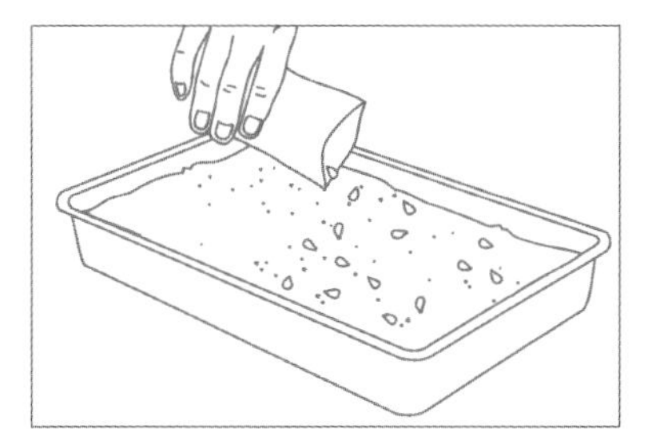

STEP 2: Sprinkle the seeds lightly in one direction and then turn the tray 90 degrees and sow the remaining seeds evenly across the surface.

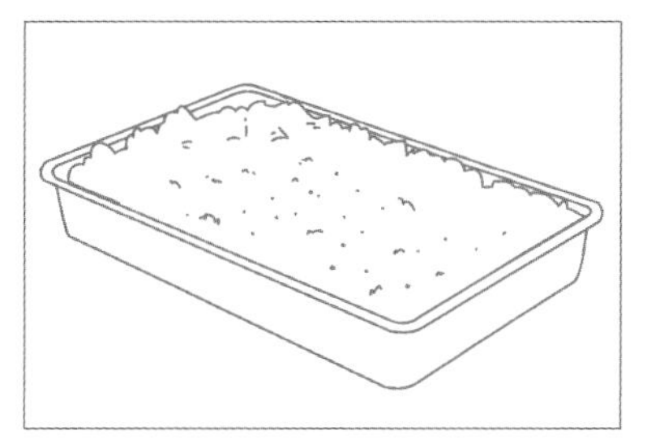

STEP 3: Cover the seeds very lightly with more of the potting compost so that they are just covered over.

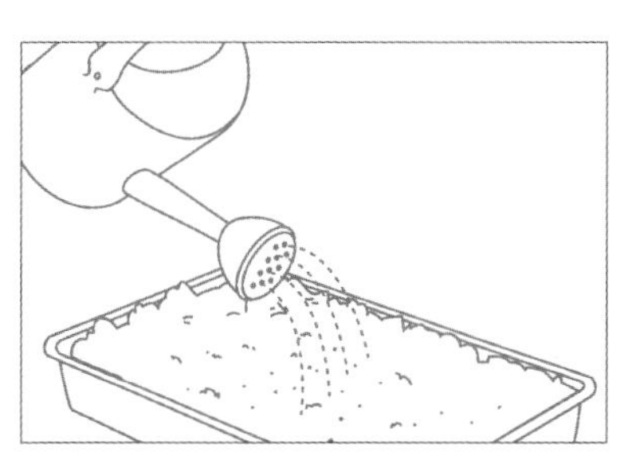

STEP 4: Water the plants in well using a watering can with a fine rose. Alternatively, stand the container in a tray of water for 15 minutes and allow to drain.

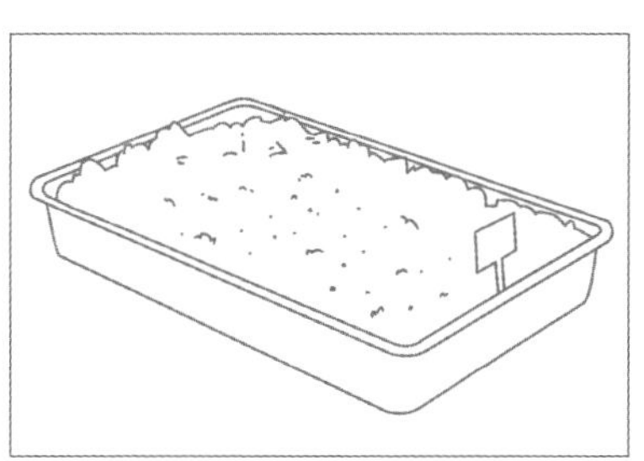

STEP 5: Remember to label the seeds with their full name and date of sowing. Place the tray on a sunny window ledge and keep the compost evenly moist but not wet.

Swiss chard

Beta vulgaris subsp. *cicla* var. *flavescens*

Common name: Spinach beet, sea kale beet, chard, perpetual spinach, silverbeet; red-stemmed types are called rhubarb, red or ruby chard; mixed colours are known as rainbow chard

Type: Annual

Climate: Half-hardy, mild winter

Size: 35cm

Origin: Sicily

History: Swiss chard does not, as its name suggests, originate in Switzerland but was named by the Swiss botanist Koch in the 19th century to distinguish chard from French spinach varieties. Its actual origins lie further south in the Mediterranean in Sicily. The ancient Greeks, and later the Romans, honoured chard for its medicinal properties not its culinary ones.

Cultivation: Sow in spring in a sunny and sheltered site in fertile soil. Leaves and stems should be ready for harvesting during summer and autumn. Alternatively, late summer sowings will provide harvests the following spring.

Storage: Like most leaf crops, they do not last for long after being picked, so harvest as needed from the vegetable plot. The stems can be chopped and frozen but will be mushy when defrosted so can only be useful for flavouring spinach-type dishes.

RIGHT: Swiss chard is a popular leafy vegetable with edible stems and foliage, and is a useful substitute for spinach, hence it common name, spinach beet.

Nutrition

Swiss chard is an excellent source of vitamins A, C and K as well as a good source of magnesium, potassium, iron and dietary fibre. It also contains phytonutrients (shown in the vibrant colours of chard) which are known to provide antioxidant, anti-inflammatory and detoxification support.

Preparation: Stems should be separated from the leaves. Young leaves simply need washing and can be added whole or chopped to brighten up salads. Do not soak leaves as this will result in loss of water-soluble nutrients to the water. Remove any brown or slimy bits of the leaves and any damage. The stalks should then be trimmed. If they are too fibrous then simply make incisions, as you would celery, near the base of the stalk and peel away the fibres. Mature chard is tougher and should be typically cooked or sautéed.

Chard is a popular leaf salad crop that comes in a range of bright colours that can brighten up the dullest of days on the allotment or vegetable plot. Both the stems and the leaves can be eaten and are popular either raw or steamed. The young leaves are particularly suitable for using in salads, whereas the more mature leaves and stems are steamed or sautéed to reduce their bitterness. Their flavour is reminiscent of cooked spinach. Leaves should be boiled or steamed for two or three minutes, stems a couple of minutes longer. Stems can also be stir-fried or even roasted. There are generally three different types of Swiss chard.

RIGHT: Chard comes in a range of stunning bright colours that look great in the garden, and brighten up many salads dishes. The mixed colours are known as rainbow chard.

Tasting notes

Cheesy chard gratin

This side dish is a quick and tasty recipe using chard leaves and stems. It can be garnished with colourful stems of rainbow or red-stemmed chard.

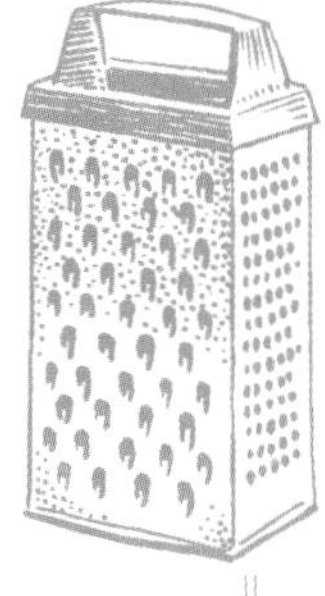

Preparation time: 10 minutes
Cooking time: 30 minutes
Serves: 6 people (as side dish)

- 340g (11oz) chard
- 150ml (¼ pint) double cream
- 1 tbsp wholegrain mustard
- 140g (5oz) strong flavoured cheese such as mature cheddar or Gruyère, coarsely grated
- 2 tbsp Parmesan, finely grated

Pre-heat a conventional oven to 200°C (400°F / gas mark 6 / fan 180°C).

Remove the leaves from the stalks and chop the stems into matchstick-thick strips.

Mix the cream, Gruyère and mustard with the chard in a gratin dish.

Grate Parmesan into the dish and place in the oven for 30 minutes.

Firstly, there are the popular brightly coloured stems known as rainbow chard, which is not a single variety but a mix of coloured types, and secondly there are the red-stemmed varieties known as ruby, red and rhubarb chard. Finally, there are the standard green glossy-leaved chards held aloft on attractive white stems. All of them add a wonderful splash of colour to the kitchen garden. Just to complicate things further, there is perpetual spinach, which is very similar to Swiss chard but has slightly thinner stems and is an excellent alternative to standard spinach. All of them are grown in exactly the same way, though chard is often preferred by gardeners as, unlike true spinach, it does not run to seed at the first hint of drought.

Chard likes a warm, sunny and sheltered site. Dig in lots of organic material in autumn before planting and sow the seeds directly into the soil in spring. Use the edge of a draw hoe to create a shallow drill about 1cm deep and sow every 40cm. Rows should also be 40cm apart. Keep the plants well watered and regularly weed between them. A late summer sowing can also be made for a spring crop, but this will need protection in cold areas with a fleece during the colder period, although in milder areas it is tough enough to survive without.

Harvest as and when required during summer and autumn. Chard is a bit like large cut-and-come-again plants, whereby stems and leaves can be harvested when needed and yet they will keep producing stems. Use a sharp knife to cut the stems at the base of the plant. They are fairly fast growing and are usually ready for picking about 10 weeks after sowing, although the sweet-flavoured baby leaves can be picked after 4 to 6 weeks.

BELOW: This historic, colourful illustration, dated pre-1400 from the *Tacuinum Sanitatis*, a Medieval health handbook, depicts a woman harvesting chard from a kitchen garden.

Beetroot

Beta vulgaris

Common name: Beetroot, table beet, garden beet, red beet, golden beet or beet

Type: Annual

Climate: Half-hardy, mild winter

Size: 35cm

Origin: Mediterranean

History: The beetroot evolved from wild sea beet, that is native along the coastlines from India to Britain. This would explain why at first it was only the leaves that were used for cooking purposes. Generally used for medicinal rather than culinary purposes it helped aid digestion and was used to cure ailments of the blood. Around 800 BC beetroot was mentioned in an Assyrian text as growing in the Hanging Gardens of Babylon and was even presented to the sun god Apollo at his temple in Delphi. The beetroot also began to appear in Roman recipes, being cooked with honey and wine, but it was not until the 18th century that the beet actually became widely used in Central and Eastern Europe, where most of the recipes used today come from.

Cultivation: Sow beetroot from early spring and harvest during summer and autumn. It prefers fertile, well-drained soil. Sow every two or three weeks if you want to harvest a continual supply of mini beets for their tender, succulent flavour.

ABOVE: Beetroot has a rich and historic horticultural past and is even thought to have been one of the vegetables growing in the legendary Hanging Gardens of Babylon.

Storage: Beetroot can be left in the ground until needed, except in very cold areas where it should be lifted and placed in trays of moist sand. Small beetroots can be pickled in jars of malt vinegar after boiling and peeling them.

Preparation: Twist off the stalks about an 2.5cm (1in) above the roots and wash the beetroot. Take care not to pierce the skin, or juices will bleed into your cooking water. Beetroot can be boiled in salted water until soft, which can take up to 1½ hours for a large beetroot, or alternatively baked in the oven at 180°C (350°F / gas mark 4 / fan 160°C) for 2–3 hours. It can be peeled and sliced and served hot in melted butter or cold in salads. If adding to salads, do this last minute or juice can bleed into the other ingredients.

For the gourmet cook there are two treats in store from just the one plant. Firstly, there are the deliciously flavoured beetroot leaves, which are brightly coloured and can be picked when very small, and then there are the juicy and succulent roots, which have a unique earthy flavour. The roots can either be picked when young and tender or left to mature and be cooked then used in a huge array of dishes. One unfortunate side effect is that not only does it stain your clothes, skin and can make you look like you are wearing lipstick, but it can even turn your toilet waste red or purple. The beetroot-red colouration comes from its sap. Growing your own enables you to enjoy beetroots that you rarely see in shops, including white and gold varieties. There are also impressive varieties containing concentric rings or white within the red flesh. 'Albinia Vereduna', for example, is sweet with white roots and sap that does not stain clothes and 'Burpee's Golden' has orange skin and yellow flesh, so it keeps its colour when cooked and does not bleed. The 'Babieto di Chioggia' variety has white internal rings when the flesh is sliced.

For a regular harvest of the roots and leaves, sowings should be made every couple of weeks from spring through to late summer. Some varieties have been bred for their winter storage qualities, meaning that it is almost possible for the vegetable garden to supply beetroot all year round.

Large beetroots are often sliced into salads and smaller ones eaten whole, but their versatile flavour can be used in so many dishes. They can be roasted as a flavoursome side dish or even used as a garnish slice in burgers. Their rich flavour is also used in sweet dishes such as chocolate brownies (see box).

Beetroots should be grown in fertile well-drained soil in full sun. They can be sown as early as February under cloches and outdoors without protection in March and April. Create a shallow 2cm drill and sow three seeds every 10cm in rows 30cm apart. For early sowings it is best to use bolt-resistant varieties. When the seedlings are about 5cm high they should be thinned to one beetroot per 10cm. Do not throw away the thinnings but eat the immature bulbs and leaves. Keep the plants well

LEFT: Beetroot is a quick-growing vegetable that has a succulent edible rootball after cooking, but also has striking edible foliage with red veins down its centre.

watered and weed free during the growing season, and harvest the beetroots when they are the size of a golf ball, leaving the ones that remain to further develop to the size of a tennis ball.

If you want to grow something different then try 'Touchstone Gold' which has bright yellow roots, an earthy flavour and is both tender and sweet. Alternatively, try a white beetroot called 'Albina Vereduna'. Not only does it have tasty white roots, but is pointy and also has leaves and stems that make a suitable substitute for chard. In fact, the plant was bred for its leaf beet qualities. The white root is a practical choice if you are worried about staining your kitchen surfaces with bright red beetroot juice – it is a far more discreet option.

BELOW: By the 18th century beetroot had become widely used in culinary dishes in Central and Eastern Europe, and this is where many of the recipes used today come from, including borscht.

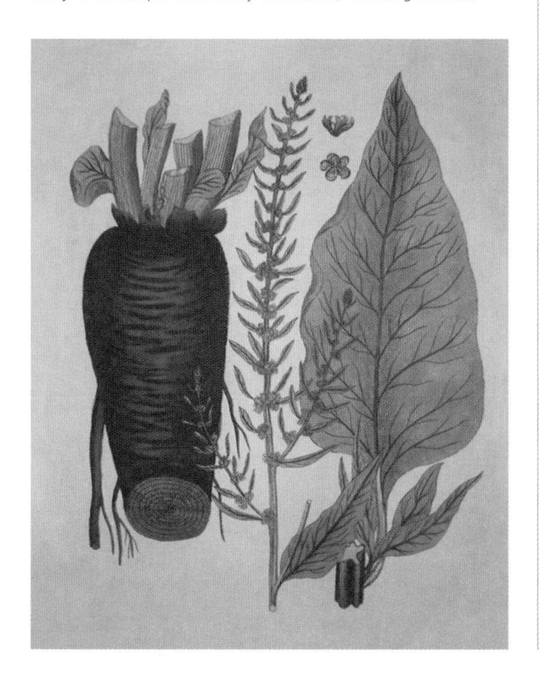

Tasting notes

Beetroot brownies

The deep red colour and earthy, rich flavours of beetroot fuse well with the sweetness and texture of chocolate in this brownie recipe.

Preparation time: 20 minutes
Cooking time: 25 minutes
Serves: makes 24 squares

- 250g (8oz) beetroot
- 250g (8oz) salted butter, cubed
- 250g (8oz) dark chocolate
- 3 medium eggs
- 250g (8oz) caster sugar
- 150g (7oz) self-raising flour

Pre-heat a conventional oven to 180°C (350°F / gas mark 4 / fan 160°C).

Boil the beetroot until tender, allow to cool, then grate.

Melt the butter and chocolate in a bowl over a pan of hot water.

Beat together the eggs and sugar, and mix in the melted butter and chocolate until smooth.

Sieve in the flour.

Add the beetroot and fold everything together with a large metal spoon.

Pour into a greased and lined shallow baking tin, about 20 x 25cm, and level with a spatula. Bake for 20–25 minutes.

Cool on a wire rack, then cut into squares.

Turnip

Brassica rapa Rapifera Group

Common name: Turnip, rutabaga, yellow turnip or neep

Type: Annual

Climate: Half-hardy, mild winter

Size: 25cm

Origin: Europe

History: Used as a vegetable in Europe since prehistoric times, the is evidence of the turnip's domestication over 4,000 years ago. A well-established crop in Roman times, the turnip was considered a food for the poorer country folk that could be grown easily and cheaply. It is now widely grown around the world.

Cultivation: Turnips prefer an open sunny but fertile soil improved with plenty of well-rotted organic matter to prevent plants drying out. Seeds should be sown in late winter for early spring harvests with regular sowings from early spring until midsummer for a continual supply as late as early winter.

Storage: Turnip greens should be eaten soon after harvesting and will only keep in the fridge for a few days. Small spring and summer turnips will keep for a couple of weeks in the fridge, whereas the larger autumn and winter turnips can be left in the ground over winter until required but in areas with mild winters only.

ABOVE: Turnips are a member of the cabbage or brassica family. They are grown for their tasty swollen roots. Their flavoursome leaves are also edible and suitable for salad.

Preparation: Cut off the greens from the top of the turnip, which can be steamed and eaten as spring greens. Young baby turnips should not need to be peeled, but larger and older types should have their outer skin removed with a knife by carving down the sides just as you would cut a pineapple. Then slice, dice or cut into chunks. Place in cold salted water and boil for 20 minutes or until tender. Drain and serve or mash lightly and serve hot with butter. Young turnips can be served raw, sliced thinly and put into salads.

For centuries, turnips were considered a staple peasant food. How times have changed, as these flavoursome root vegetables are now highly sought after – particularly the young and tender spring turnips.

Turnips, like swedes, are a member of the cabbage family and grown for their large, swollen root. Confusingly, in Scotland swedes are also called 'turnips', or 'neeps' for short.

Most turnips are perfectly round but there are more flattened forms. Skins are usually creamy white and their tops are usually green, purple, white or yellow. The flesh is usually white or yellowish. The vegetable is mainly grown for the swollen root, which can be eaten when small and succulent in spring or summer. The winter types are larger, tougher and often cooked in casseroles, stews and soups. The young leaf tops are also a gourmet treat and have a slightly peppery flavour. They are cooked and eaten as spring greens.

As turnips are part of the cabbage family, they are best grown with other brassicas, such as kale, Brussels sprouts and cauliflower. In this manner, all their needs can be met as if growing one crop.

Like most root crops, turnips do not transplant well, and so they should be sown directly where they are to be grown. Early varieties can be sown under cloches in late winter to produce the young succulent roots for spring. From early spring onwards, they can be sown directly into the soil without the need for protection. Create a shallow drill and sow them in rows 20cm apart. Seeds should be sown into 1cm-deep drills, and the seedlings thinned out to 12cm apart when they are about 5cm tall.

For autumn and winter harvesting, seeds should be sown in midsummer, but in rows that are 30cm apart as these roots will get much bigger. The seedlings should be thinned out to 20cm apart.

'On rainy days he sat and talked for hours together with his mother about turnips ...'

Mark Twain, *Roughing It*, (1886)

RIGHT: A watercolour on paper of a turnip from *Collection du Regne Vegetal, Fleurs, Plantes, Arbres, et Arbustes* possibly by Pierre Francois Ledoulx, a Belgian painter of flowers and insects.

Harvest the late winter, spring and early summer sowings when they are about the size of golf balls for the best flavour. Sow every two or three weeks in short rows, for a continual supply of tender turnips. Keep the plants well watered to ensure they grow quickly and keep them weed free as they grow.

Midsummer sowings should be harvested from autumn onwards. They can be left in the ground until required, but in very cold areas where the ground is likely to be frozen solid over winter, it is better to dig them up with a fork, remove the leaves and store them in damp sand.

The old Celtic festival of Samhain (Hallowe'en) traditionally carved candle lanterns out of turnips and not pumpkins.

BELOW: *The Incorruptible Consul Curius Dentatus Preferring Turnips to Gold* (1656). Turnips have a long, colourful history of cultivation and were often the subject of stories and paintings.

Tasting notes

Mashed turnips with crispy bacon

Mashed turnip is just as smooth and creamy as mashed potato, but with a slightly stronger flavour. The added crunch of crispy bacon makes this a great side dish to meat dishes, particularly roast lamb, beef or haggis.

Preparation time: 20 minutes
Cooking time: 30 minutes
Serves: 4 people

- 3–4 turnips
- 60g (2oz) butter
- Salt and pepper, to taste
- 120g (4oz) cooked bacon, chopped
- Handful of chives, chopped
- 50g (2oz) Parmesan, grated

Pre-heat a conventional oven to 200°C (400°F / gas mark 6 / fan 180°C).

Peel and cut up the turnips, then boil in salted water until tender.

Drain and mash with butter, then season.

Fold in the cooked bacon and chives.

Cover with Parmesan and cook in the oven for 30 minutes.

(For a stronger flavour, add goat's cheese, or blue cheese such as Stilton, Roquefort, Gorgonzola, Cambozola or Danish Blue.)

Swede

Brassica napus Napobrassica Group

Common name: Swede, yellow turnip, Swedish turnip and Russian turnip and, in America, rutabaga. In Scotland, swedes are called neeps

Type: Annual

Climate: Hardy, average to cold winter

Size: 30cm

Origin: Central Europe

History: The swede is thought to have originated in Central Europe and has a relatively short culinary history compared with many vegetables. In 1620, a Swiss botanist called Gaspard Bauhin noted that swede was growing wild in Sweden, which is where the name came from, and it was known in France and England during this century; it is recorded as being present in the Royal gardens of England as early as 1669. By the 18th century, it had become an important European crop.

Cultivation: Sow the seeds in fertile sunny sites, enriched with organic matter, in spring. Harvest in autumn and late winter.

Storage: Leave swedes in the ground until required or dig them up and store in sand in a garage or cellar if there is a risk that they may become frozen in the ground.

ABOVE: Swedes are a popular root vegetable, belonging to the cabbage family. Its first recorded use as a vegetable is from the 17th century.

Preparation: Peel thickly to remove all the skin and roots before use. As the skin is quite thick and uneven, you may find it easier to quarter the swede first and then cut off the skin with a knife. Cut swedes into chunks or cubes, according to preference, and place in cold salted water and boil for 20 minutes or until tender. If roasting them, place in the oven at 200°C for around 30 to 45 minutes. Swede can also be used raw and is delicious finely grated and tossed into a salad.

Closely related to the turnip, the swede is a welcome treat in the winter vegetable garden when there is little else to harvest. The flavour is milder and sweeter than the turnip and it tastes sensational when roasted or sautéed. Mashing it with garlic and cream is the perfect accompaniment to a roast meal, while it can also be added to bulk up casseroles and soups. Due to its mildness it benefits from plenty of seasoning, particularly pepper, to get the best from its subtle flavour. When cooked it turns a light orange colour, that looks attractive on the plate, particularly when contrasted with other winter vegetables such as leeks and kale.

Unlike its close cousin the turnip that is sown throughout the year for frequent harvesting, swedes only have one season of interest in winter, so enjoy it while it lasts.

Like other members of the cabbage family, swede requires a fertile soil that has been enriched with organic matter to help retain moisture. It prefers a sheltered site in full sun. On acid soil swedes can be prone to a disease called clubroot, which deforms the root system. To avoid this, it is beneficial to add a dressing of lime before planting to increase the alkalinity.

Like most root crops, swedes do not transplant well and should be directly sown in spring into shallow 1cm drills, in rows 35cm apart. The seedlings should be gradually thinned out until they are at a final spacing of 25cm apart in the row. Keep the plants well watered during dry periods to avoid the roots splitting. It is also necessary to remove weeds regularly to prevent them competing for moisture and nutrients.

Swedes are ready to harvest from autumn onwards, but there is no need to harvest them immediately as they are tough as old boots and will remain out in the winter cold until required in the kitchen. As with turnips, in really cold areas where the ground may freeze over, they can be dug up in late autumn and stored in damp sand in a garage or cellar.

Gaspard Bauhin

A Swiss botanist and anatomist, Gaspard Bauhin was born in Basel, the son of a French physician, and studied medicine at Padua, Montpellier, and in Germany. He was also a pioneer in binomial nomenclature and organized the names and synonyms of 6,000 species in his illustrated exposition of plants – *Pinax theatri botanici*, which has become a landmark of botanical history.

This classification system was quite basic and used traditional groups such as trees, shrubs and herbs. He did, however, correctly group legumes, grasses and several others. His most important contribution is in the description of genera and species. He introduced many names of genera that were adopted by Carl Linnaeus, a Swedish botanist, and remain in use today.

'This is the plant which the English Government thought of value enough to be procured at public expense from Sweden, cultivated and dispersed. It has such advantage over the common turnep that it is spreading rapidly over England and will become their chief turnep.'

Thomas Jefferson, letter to a friend, (June 1795)

BELOW: The swede is an invaluable vegetable during the colder winter months when there is not much else available to harvest from the kitchen garden. If left to go to seed, its yellow flowers are attractive.

Tasting notes

Spicy swede wedges

This is a dish that puts the spice into one of the more traditional vegetables, the swede. These wedges make a great starter – try them dipped in a curry sauce, or as a side dish to traditional meals such as a Sunday roast.

Preparation time: 5 minutes
Cooking time: 30–35 minutes
Serves: 4 people

- 750g (1lb) swede, peeled
- 1 tbsp olive oil
- Pinch of paprika
- Salt and pepper, to taste

Pre-heat a conventional oven to 200°C (400°F / gas mark 6 / fan 180°C).

Slice the swede into finger-width discs. Slice again across each disc to make thin wedges.

Tip the swede, olive oil and paprika into a shallow roasting tin. Season and toss well, arranging in one layer.

Roast for 30–35 minutes, turning half way through cooking, until golden brown.

Drain on a kitchen towel and lightly salt to serve.

Kale

Brassica oleracea Acephala Group

Common name: Kale, borecole

Type: Annual or Biennial

Climate: Hardy, cold winter

Size: 35cm

Origin: Asia, Mediterranean

History: Kale has been cultivated for over 2,000 years and is a descendent of the wild cabbage, a plant thought to have been brought to Europe around 600 BC by groups of Celtic wanderers. In much of Europe it was the most widely eaten green vegetable until the Middle Ages when cabbages became more popular.

During World War Two, the cultivation of kale in the UK was encouraged by the 'Dig for Victory' campaign. The vegetable was easy to grow and so provided important nutrients to supplement those missing from a normal diet because of food rationing.

Cultivation: Kale should be sown directly outdoors into shallow drills. When they reach a height of 10cm they can be transplanted to their final planting position, 45cm apart. Leaves are ready for harvesting from autumn until spring.

Storage: Kale is winter hardy so can remain in the ground until needed in the kitchen.

ABOVE: Kale is a member of the cabbage family grown for its healthy leaves. It makes an attractive addition to the garden during the winter months with its strong structure and texture.

Once picked, cook it within a day or two. Alternatively, cook it in dishes and freeze for eating later.

Preparation: Snap off the stalks; wash the kale thoroughly in cold salted water and drain. The leaves can be cooked whole or chopped up.

Kale leaves are enjoying a bit of a renaissance among chefs and gourmets, who admire the versatility of this leafy member of the cabbage family. It goes well with fish and meat, and is usually boiled or steamed but is equally good when used in stir-fries and casseroles. Some people may find the flavour slightly bitter, but when cooked well it provides a wonderful background flavour in soups and stews. Combine with cheese, onion and eggs they make for a delicious winter warming filo pastry pie.

Gardeners also appreciate this previously underrated vegetable for a number of reasons. Firstly, it is fully winter hardy and fills a gap in the

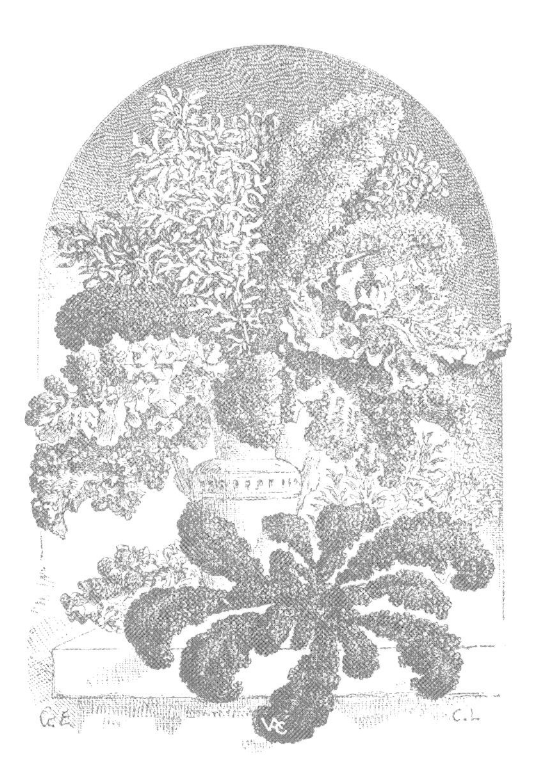

ABOVE: There are lots of different varieties of kale but they all have attractive foliage ranging from crinkly red-leaved varieties through to dark narrow-leaved types such as 'Nero di Toscana'.

Tasting Notes

Kale with roasted peppers and olives

This quick and easy kale recipe makes for a very healthy snack or side dish.

Preparation time: 10 minutes
Cooking time: 25 minutes
Serves: 4 people

- 2 large bunches kale
- 2 tbsp olive oil
- 2 cloves garlic, thinly sliced
- 60ml (2fl oz) water
- 2 tsp sugar
- 1 tsp salt
- 12 Kalamata olives, pitted and chopped
- 120g (4oz) jar of roasted red peppers
- 2 tbsp aged balsamic vinegar

Cut the kale into bite-size pieces, removing any tough stems. Rinse and shake dry.

Heat the oil and brown the garlic in a pan.

Add the kale and stir-fry for 5 minutes.

Pour in the water, cover, and cook for 8–10 minutes, or until tender.

Spoon in the sugar, salt, olives and peppers.

Cook over medium-high heat until the liquid has evaporated.

Plate up; scatter the garlic over the top and drizzle with balsamic vinegar.

late winter and early spring culinary calendar when there is little else available on the plot. Secondly, it is more tolerant than most of the other brassica plants to poor soil and wet conditions. Kale is also valued for its fantastic-tasting and nutritionally packed leaves, and in the garden the different colours and forms can provide a beautiful tapestry of textures. In fact, they give a wonderful display when used to edge borders, which is unsurpassed by other ornamental plants in the depth of winter. Varieties such as 'Red Russian' have attractive crinkly red leaves, while the dark narrow-leaved variety 'Nero di Toscana' (sometimes called palm tree cabbage) provides an attractive visual depth to any vegetable bed.

BELOW: Kale is grown for its tasty foliage and there are many ornamental varieties to choose from. These can also be used to provide evergreen structure in herbaceous borders and flower beds.

Kale requires a fertile soil in full sun. Prior to sowing, the site should be thoroughly dug over and lots of organic matter added such as garden compost or well rotted manure. Seed should be sown in late spring either in pots or directly outside into shallow drills that are 1cm deep. Rows should be 20cm apart. When they reach a height of about 10cm high they can be transplanted to their final planting position, at spacings of 45–60cm, depending on the variety, in rows 60cm apart. Keep the plants well watered during summer, and weed between the rows and around the plants each week to prevent any competition for nutrients and water from these hungry vegetable plants.

Kale is extremely hardy and the plants will remain resolutely outside in the freezing cold weather. When harvesting, it is best just to take a few leaves from each plant as required rather than stripping an entire plant in one hit as it may not recover. Harvest the lower leaves first, before using the leaves higher up later in the season.

Nutrition

Kale is high in iron which is essential for good health, since iron is used in the formation of haemoglobin to transport oxygen to various parts of the body, cell growth, and liver function. It is also high in vitamins A, C and K, which help to maintain a healthy body and immune system. It is also recommended for detoxing as it is filled with fibre and sulphur.

Cabbage

Brassica oleracea Capitata Group

Common name: Cabbage, cabbage leaf, green cabbage

Type: Annual

Climate: Hardy, average to cold winter

Size: 40cm

Origin: Europe

History: The word 'cabbage' is an Anglicized form of the French *caboche*, meaning 'head', referring to its round bulbous shape. In addition the word *Brassica* comes from the Celtic word *bresic*, meaning 'cabbage'. Cabbage has been cultivated for more than 4,000 years and domesticated for over 2,500 years. Since cabbage grows well in cool climates, yields large harvests, and stores well during winter, it soon became a major crop in northern Europe.

Cultivation: Cabbages require a fertile soil with plenty of added organic matter in full sun. Sow in modules or in nursery beds to transplant into its final position later. Sowing times depend on when the cabbage is to be harvested during the year.

Storage: Most cabbages are hardy and can remain in place until required in the kitchen. On harvest, they can be stored in a cool place for several weeks or longer, depending on type.

ABOVE: The illustration above is a chromolithograph of cabbage varieties taken from the *Album Benary*, illustrated by Ernst Benary, and dates from 1876. The album contains 28 colour plates of different vegetable varieties, named in English, German, French and Russian.

Preparation: Remove the outer leaves first and cut the cabbage in half. Cut out and discard the centre stalk, then wash and cut the leaves as required. Cabbage can be shredded for using raw in salads; for cooking it can be cut into thick wedges; alternatively the centre can be stuffed. Shredded red cabbage is best braised.

Where would the culinary world be without cabbages? Much better off, might think many people who have lingering memories of overcooked, boiled cabbage for school dinner. However, there are many fantastic reasons why cabbages are one of the most widely grown vegetables, as they have been for centuries. There are so many famous international dishes made from this staple vegetable, and possibly two of the best known are the fermented sauerkraut and the popular salad coleslaw, where it is thinly chopped up with other raw vegetables and mixed with mayonnaise. However, it is a wonderful hearty vegetable and has so many more uses in the kitchen. What can be more satisfying than a large bowl of mashed potatoes with steamed cabbage, garlic, onions and Bramley apples in the depth of winter?

Cabbages look beautiful in the vegetable garden and create a wonderful tapestry of texture and colour. They come in many shades including green, red and purple, and in a range of different shapes from pointy to spherical or open. They can be smooth or crinkly and provid structure in the kitchen garden throughout the year if planted in blocks.

There are cabbages for every season and are so named after the period they are picked: summer, autumn, winter and spring. These are mainly hearting cabbages, but spring cabbages can also provide loose heads of green leaves. Savoy cabbages are a popular type of winter cabbage as are some of the white-, red- or purple-tinted varieties, one of the most popular being 'January King'.

The cultivation of cabbages, whatever the season is basically the same; the main difference is when they are harvested. Cabbages should be grown in full sun in fertile, well-drained soil. They are hungry plants so enrich the soil with plenty of organic matter prior to sowing. Most gardeners raise the seedlings in outdoor nursery beds before transplanting, but they can be grown in modules indoors or in situ if space allows.

Tasting notes

Creamy coleslaw

Coleslaw is a classic cabbage recipe and is now a popular side dish and garnish around the world. For a more healthy alternative, use natural yogurt instead of mayonnaise to coat the delicious shredded vegetables.

Preparation time: 10 minutes
Serves: 10 people

- 1/2 white cabbage, finely shredded
- 2 large carrots, finely shredded
- 6–8 tbsp mayonnaise
- 1/2 red onion, thinly sliced
- 1 tsp lemon juice
- Salt and pepper, to taste

Combine the shredded cabbage and carrots in a large bowl.

Stir together the mayonnaise, onion, lemon juice, salt and pepper in a medium bowl.

Add to the cabbage mixture. Mix well. Serve.

RIGHT: Cabbages provide a beautiful splash of colour in the garden all year round particularly throughout the winter months. They have a variety of textures, shapes and colours as can be seen with the facing illustration of Savoy, red and white cabbages.

WISE WATERING

Water is the giver of life. Without it, plants cannot survive and in dry summers this can cause a real problem. However, it is important to get it right as plants can also die from being over-watered.

Water during the summer is becoming a rare commodity in certain areas and so it is important that plants are watered responsibly and with minimum wastage.

LEFT: Watering cans are the best way to moderate and control the amount of water to be used. Roses attached to the end of the nozzle distribute a fine spray.

TIPS FOR RESPONSIBLE WATERING

Avoid using sprinklers as they tend to waste a lot of water and do not always provide plants directly with water at their roots. Seep hoses are more effective as they target the root area, slowly soaking water into the soil and reducing the amount of evaporation. Seep hoses can be bought from most garden retail outlets, but can simply be made by poking small holes in a standard hose with a sharp needle. Lay the hose down among the plants.

Using a watering can gives better directional control than using a hose, thereby avoiding wastage. Ideally it should be aimed at the area directly surrounding the root zone of the individual plant, avoiding the leaves. Allow the water to soak through thoroughly, placing a rose over the end of the nozzle slows down the water flow, giving it time to soak into the soil and avoid run-off. Roses are also the best way to water delicate seedlings. Creating a sump or hollow around individual plants also helps to direct water to where it is needed.

Watering is most effective in early morning or in the evening. Avoid watering during the middle of the day, particularly in warm weather as much of the water will be lost to evaporation.

Use shade netting or shade paint in the glasshouse as this will reduce water loss and prevent the leaves from getting scorched.

Plants grown in containers and hanging baskets usually require more watering than those grown in the ground. However, their watering requirements can be reduced if they are moved into the shade during the hottest part of the day. Set up water butts and any other water-tight collection units so that rainwater can be harvested. Attach them to the down pipes on glasshouses,

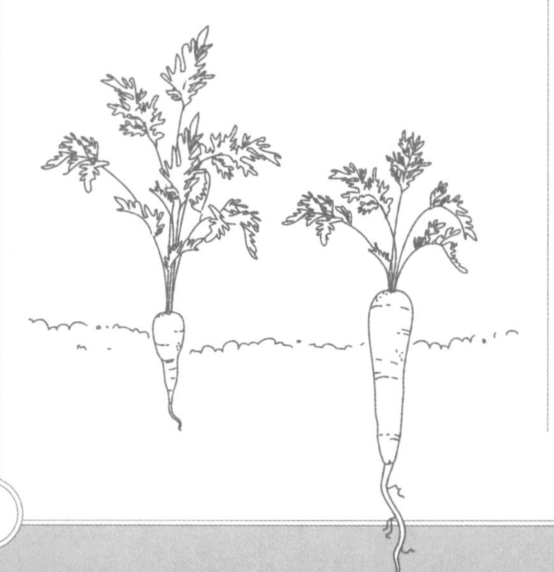

LEFT: Excessive watering and fertilizing can lead to lots of foliage (left) at the expense of root development. Regular but not excessive watering is better.

When to water what

The general rule of thumb is that plants need regular watering during dry spells, but individual types of vegetables do have varying requirements.

Potatoes	Regular watering is especially important during the flowering period as this is when the tubers start to form in the ground. Give them one good drenching every few days, rather than each day as little and often encourages shallow rooting.
Carrots	A free-draining soil is preferred, and therefore they should not be kept too moist. Over-watering will lead to an excess of leaves and a stunted root system, as it does not need to grow deeply to find moisture.
Onions	Avoid over-watering onions when the bulb is swollen as this will not help them cure and harden prior to harvesting. Excessive watering can also lead to fungal problems. Pulling back some of the moisture-retaining mulch will also expose the bulb to the sun and help keep it drier.
Cabbages	Give cabbages a thorough soaking when planting and they should only need regular watering every few days afterwards. Once their heads start to form, generous amounts of watering will improve their size.

ABOVE: Peas need plenty of moisture, particularly around flowering time to encourage pods to form.

garages, sheds and the house to maximize the amount of water that can be collected.

Recycle water from baths, showers and the kitchen sink which is sometimes called grey water. Avoid reusing water that is contaminated with too much soap, or detergents and bleach.

Containers such as plastic cartons and bottles can have their bases removed and be buried mouth-downwards next to individual plants. When topped up with water, they will slowly seep out water into the root zone, where it will be most needed.

Mulch vegetable beds with garden compost or well-rotted manure; this will help to keep moisture in the soil. Mulches also suppress weeds, which will compete with your vegetables for moisture and nutrients.

Cauliflower

Brassica oleracea Botrytis Group

Common name: Cauliflower

Type: Annual

Climate: Half-hardy to hardy; mild to cold winter

Size: 35cm

Origin: Mediterranean

History: Cauliflower can trace its ancestry to the wild cabbage and has been an important vegetable in Turkey and Italy since at least 600 BC. It became popular in France in the mid-16th century and was subsequently cultivated in Northern Europe and the British Isles. The United States, France, Italy, India, and China are countries that produce significant amounts of cauliflower.

ABOVE: The beautiful creamy white florets of the cauliflower plant makes this one of the most attractive members of the cabbage family but also one of the trickier ones to grow.

Cultivation: Cauliflowers prefer a fertile, well-drained soil in sun. Seeds can be started outdoors by sowing in shallow drills 1cm deep in rows 20cm apart and transplanted to their final position of 50cm apart. Sowing times vary depending on variety and season of harvest.

Storage: Cauliflowers will keep better if they are stored upside down in a cool dark place. They can also be preserved as pickles and preserves, the most popular being piccalilli. Florets can be placed in containers and frozen.

Preparation: First, simply cut away the outer leaves and chop off the stem. You can also cook with some of the most tender leaves that are found around the cauliflower so do not discard completely. Next chop the head into pieces and cut a cross in the stump of each to aid in the cooking process. Wash and drain, then either boil or steam. It can also be served raw in salads or dipped in batter and deep-fried.

Sometimes the simplest food is the best, and there is nothing simpler or better than one of the nation's favourite: cauliflower cheese (see box).

To be fair, cauliflowers are probably one of the trickier members of the cabbage family to grow, but with them being expensive in the shops and a wider range of attractive and flavoursome varieties to try from the seed companies, they are well worth the effort.

There are plenty of different varieties to choose from including orange-yellow ones, 'Cheddar', and ones with deep purple heads, 'Graffiti'. However, even the usual white-headed dome varieties look attractive in the flower garden. Most cauliflowers produce large curds – the gardener's name for the dome-shaped flowerheads – which usually measure about 15 to 20cm across. However, the modern hybrid mini cauliflowers producing curds just 10cm across are now becoming increasingly popular with chefs and gardeners alike.

BELOW: This still life painting called 'Egg and Cauliflower' painted by George Washington Lambert in 1926 clearly illustrates the attractive and textural qualities of this commonly grown vegetable.

TASTING NOTES

Cauliflower cheese

Cauliflower cheese is easy to prepare and versatile. It is filling enough to be served as a main meal or as a side dish to a roast dinner.

Preparation time: 5 minutes
Cooking time: 30 minutes
Serves: 6 people (as a side dish)

- 75g (2½oz) butter
- 50g (2oz) plain flour
- 1l (1¾ pints) milk
- 100g (4oz) cheddar cheese, grated (plus 25g/1oz grated for topping)
- 1 medium cauliflower, cooked florets
- 4 slices white bread, breadcrumbs
- 75ml (2fl oz) double cream

Melt three-quarter of the butter in a saucepan and stir in the flour. Cook for 1 minute.

Gradually add the milk, stirring to a smooth sauce. Simmer for 15–20 minutes until thick, then sprinkle in half of the cheese.

Melt most of the remaining butter in a pan and fry the cauliflower until slightly browned. Spoon into a baking dish.

Put the rest of the butter into the pan, add the breadcrumbs and fry until golden.

Stir the remaining cheese into the sauce until it melts and add the cream.

Pour the sauce over cooked cauliflower florets.

Cauliflowers can be produced almost all year round by choosing the correct varieties and sowing them at the right time. However, winter types are technically not ready until springtime, and if you are short on growing areas in the veggie patch then they are probably not worth it as they take up so much space.

The skill with growing a cauliflower is to get the curd to develop regularly and not be misshapen. This means regular watering throughout the growing season. Very hot summers can cause problems as cauliflowers prefer slightly cooler conditions than many other vegetables.

They prefer a fertile, well-drained soil. Well-rotted manure or garden compost should be dug into the soil prior to planting as this not only enriches the soil but more importantly will help to retain the moisture and prevent the plant from drying out; a common cause of misshapen cauliflowers. Avoid acidic soil as this can cause club root, a soil-borne fungus that causes a misshapen root system. A dressing of lime can be added to temporarily increase the alkalinity of the soil.

Get seeds started outdoors by sowing them in shallow drills 1cm deep in rows 20cm apart either in a prepared seedbed or in trays of potting compost. Seedlings should be thinned to 5cm apart. Once they have produced about five or six leaves they should be transplanted to their final position and planted about 50cm apart. Sowing times vary depending on variety and season of harvest. Check the seed packet for details.

ABOVE: Historic cauliflower varieties from 1904. 'Easter Winter', 'Scilly Black' and 'Chalon Early' by Vilmorin-Andrieux. None of these are now commerically available.

During the growing season the leaves should be folded over the curd to prevent it discolouring in the sun. Plants should be covered with a protective mesh or fleece to prevent attack from birds and insects.

The summer and autumn varieties are usually ready about 16 weeks from sowing in spring. Winter varieties are ready about 40 weeks from sowing. Cut through the base of the stem with a sharp knife as and when they are needed in the kitchen. They will store for about three weeks if stored in a dark cool place.

LEFT: Cauliflowers are usually eaten for their attractive central florets but the luscious green rosettes of leaves are just as tasty as other cabbage varieties when boiled or steamed.

'Training is everything. A peach was once a bitter almond; a cauliflower is nothing but a cabbage with a college education.'

Mark Twain, *Pudd'nhead Wilson*, (1894)

Pak choi

Brassica oleracea Chinensis Group

Common name: Pak choi, bok choi, Chinese chard, Chinese mustard, celery mustard, Chinese cabbage, spoon cabbage

Type: Annual

Climate: Half-hardy, mild winter

Size: 30cm

Origin: South China

History: Records show that pak choi was first cultivated in South China as far back as the 5th century AD. By the 19th century pak choi plantations were found in Japan and south Malaya. Pak choi was not introduced to Europe until the mid-18th century when in 1751, Pehr Osbeck (see p.74), a friend of the famed botanist Carl Linnaeus, brought seeds of the vegetable to Europe, making its cultivation very popular.

Cultivation: Seeds can be sown under glass in early spring or directly in the open in mid-spring in shallow drills. Make sowings every few weeks throughout summer to regularly harvest either the baby leaves, or the semi-mature or full-headed plants.

Storage: The leaves and mature heads should be used fresh from the garden. They will only keep for a few days in the fridge, although they will last for much longer if made into soups and other dishes and frozen.

Preparation: Wash thoroughly. The leaves can be cut from the stems, as they cook at different speeds – the leaves cook much quicker, so you could add for 2 minutes just towards the end of cooking. Alternatively, to put leaves and stems in the pan at the same time, cut the stems into thin strips. Very young pak choi can be left whole; larger plants can be halved or quartered.

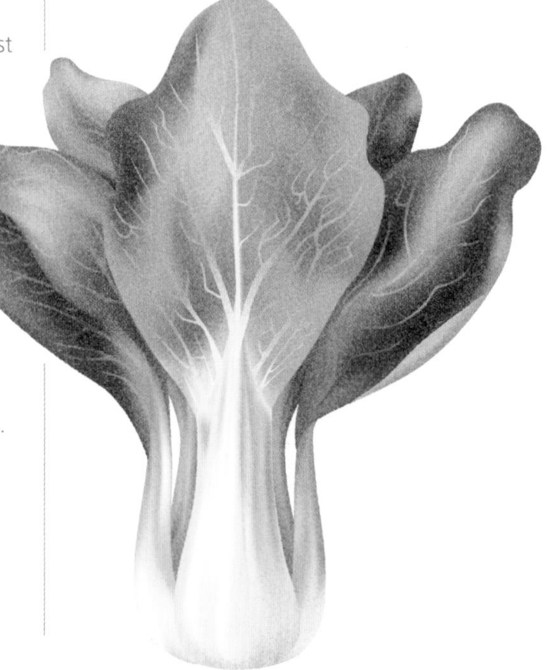

RIGHT: Pak choi is also known as bok choi and is a member of the cabbage family, grown for its tasty foliage. It is often used in Asian cuisine, particularly in Chinese and Japanese recipes.

Pak choi is also confusingly known as bok choi, but it is exactly the same vegetable. It is part of the ever-growing popularity among chefs and gardeners for a taste of the exotic – valued for its spicy, flavoursome leaves from the Far East. Seed catalogues and even supermarket shelves have a range of exciting Oriental vegetables to try including Chinese lettuce, Japanese mustard spinach (also known as komatsuna), Japanese turnip and chopsuey greens. All these vegetables have opened up a whole new culinary world not just as mature vegetables but also when grown as micro greens (baby leaf vegetables eaten young as seedlings).

Pak choi is one of the most popular vegetables in Asian cuisine. It is used in salads and stir-fries when grown for its micro leaves, but it is also used in a range of oriental dishes when it is allowed more time to develop its fully grown head. Leaves can be eaten raw or can be lightly steamed.

For those of you impatient to harvest your first crop, then the good news is that it takes merely 30 days from sowing to the picking of its first leaves. However, it takes approximately 45 to 80 days if the semi-mature or full-sized head is required, depending on the weather. As with most members of the cabbage family, pak choi requires a sunny position in fertile soil. For those people with limited space it can also be grown in a window box, making it easy to harvest the leaves for a salad whenever required. Seeds can be sown any time from mid-spring through to midsummer; sow them thinly at about 1cm deep. Rows should be about 30cm apart if the plants are being grown for their young, baby leaves, but slightly wider if they are going to develop into mature plants with

Tasting Notes

Stir-fried pak choi

This recipe is healthy and can be made as spicy as you like by adding more chilli. It is an ideal accompaniment to a bed of rice or mixed in together with egg noodles in a stir-fry.

Preparation time: 5 minutes
Cooking time: 5 minutes
Serves: 2 people (as a side dish)

- 2 tbsp sunflower oil
- 4cm (1½in or thumb length) root ginger, peeled and finely chopped
- 1–2 red chillies, finely sliced
- 3 cloves of garlic, chopped finely
- 2 pak choi, stalks finely sliced, leaves roughly sliced
- Salt, to taste
- ½ tsp soy sauce
- ½ tsp sesame oil

Heat the sunflower oil in a large frying pan or wok until very hot. Add the ginger, followed by the chilli and garlic.

Then immediately add the pak choi stalks and quickly stir.

Cook for 1 minute then add the leaves and stir until just wilting; takes about 1 minute. Then remove from heat.

Add the salt, a shake of soy sauce and a few drops of sesame oil and serve.

full-sized heads. In mild areas it is possible to sow them under cloches or fleece for an early spring start, or to extend the season into autumn.

Seedlings should be thinned out to 6cm apart if growing for baby leaves and about 30cm apart for the full-sized heads. Instead of throwing the thinnings on the compost heap, they can be eaten as delicious micro greens.

Pehr Osbeck

Pehr Osbeck was a Swedish explorer and naturalist. He was born in 1723 and studied at Uppsala, Sweden, with botanist Carl Linnaeus. In 1750, he travelled to Asia and spent four months studying the flora, fauna and people of Canton in China. His studies led him to contribute more than 600 species of plant to Linnaeus' *Species Plantarum* (*The Species of Plants*) in 1753.

ABOVE: Pak Choi will need watering most days during dry periods as they are prone to bolting to seed. Fresh baby leaves can be picked just a few weeks after sowing or the plants can be left to develop a head. The baby plants can be removed and eaten, and the remaining plants left at 30cm spacings to mature.

As the plants continue to grow it is important that they are watered every few days as in warm or dry weather they can be prone to bolting (quickly growing to seed, making the leaves taste bitter) in warm or dry weather. 'Joy Choi' is a good variety to try as it does have some bolting resistance. Other popular varieties include the dark-leaved 'Baraku', 'Choko', 'Glacier', 'Ivory', 'Red Choi' and 'Summer Breeze'.

The leaves can be picked at any stage three to four weeks after sowing. Pick them young as the older leaves can become tough. The mature heads can be harvested by cutting through the stem with a knife. Leave the stump in place, though, as it should re-sprout more baby leaves for harvesting a few weeks later. The flowerhead can also be harvested and cooked in stir-fries.

Brussels sprout

Brassica oleracea Gemmifera Group

Common name: Brussels sprout, cruciferous, sprout

Type: Annual

Climate: Hardy, average to cold winter

Size: 60cm

Origin: Belgium

History: Sprouts were believed to have been cultivated in Italy in Roman times, and possibly as early as the 1200s in Belgium. Records show that they were cultivated in large quantities in Belgium as early as 1587 (hence the name 'Brussels' sprouts). They remained a local crop in this area until their use spread across Europe during World War One. Brussels sprouts are now cultivated throughout Europe and the United States. California is the centre for Brussels sprout production in North America.

ABOVE: The humble Brussels sprout is a commonly used vegetable during the winter months due to its ability to withstand very cold conditions.

Cultivation: Sow seeds in drills in spring into fertile soil in full sun. Plants should eventually be about 50 to 75cm apart and may need to be supported with a stake to prevent them blowing over in the wind. Pick from autumn onwards and throughout winter.

Storage: The best way to store sprouts is to keep them on the plant until they are needed in the kitchen. They are fully winter hardy and so will be fine even in the depth of winter. Once picked they will keep in the fridge for a couple of weeks.

Preparation: Remove any damaged and wilted leaves and cut off the stems. If the sprouts are large some people believe it is best to cut a cross in the base to allow the thick part to cook at the same time as the leaves. Wash the sprouts to remove any dirt and pests. Brussels sprouts may be boiled in salted water for 8–10 minutes or steamed for about 15 minutes until tender.

'Brussels sprouts are a winter vegetable of great worth, each stem a column of close-packed and firm buttons with a small loose cabbage or bunch of greens at the top.'

Charles Boff, *How to Grow and Produce Your Own Food*, (1946)

Brussels sprouts must be the most maligned vegetable in the world, with many people associating them with hard, bitter bullets that have to be suffered just once a year at Christmas dinner. However, with a bit of imagination they can be cooked in many different ways, including steamed, stir-fried or added to numerous dishes, meaning there should always be some way of enjoying this very healthy and nutritional stalwart of the vegetable kingdom. Depending on variety, they can be picked from as early as August, although their flavour is considered to be better if they have been sweetened by the autumn frosts. Some sprouts will last well into late winter and early spring. At the end of the season their leafy 'sprout tops' can be removed and cooked.

When in crop, Brussels sprout plants can be top heavy. Some gardeners recommend not digging over the ground immediately prior to planting as this loosens the soil, meaning that sprouts are more prone to toppling over in the wind. However, plenty of organic matter should be added to the soil in the autumn before planting as Brussels sprouts, like other members of the cabbage family, are hungry plants requiring a fertile and moist but well-drained soil in full sun. Ideally the soil should not be too acidic, and if this is the case, a dressing of lime can be added to make the pH between 6.5 and 7.

BELOW: To cross or not to cross – in the sprout world chefs are divided as to whether or not cutting a cross through the base of the sprout is beneficial to the cooking process.

Tasting notes

Brussels sprout alternatives

If you do not like Brussels sprouts then blame Belgium as it is supposed to have originated from there, hence the name 'Brussels'.

There are now modern varieties that are sweeter than the traditional varieties such as 'Trafalgar', which is said to have reduced some of the bitterness. Alternatively there are interesting crosses between kale and Brussels sprouts such as *Brassica* 'Petit Posy Mix' that produce rosettes of loose frilly-edged buttons in an attractive mix of purple and green colours. The flavour is closer to eating spring greens, but it is a good substitute for those who feel that there should be a sprout-related vegetable on the Christmas dinner plate. Finally, if you really cannot face a side dish of sprouts on their own, there are lots of delicious recipes (see p.77).

Sprouts can be sown indoors in modules in February and planted out in spring. Alternatively, they can be sown directly outdoors into a nursery or seed bed in March ready for planting in their final position later.

Sprouts are usually ready for transplanting into their final position about five weeks after sowing or when the seedlings have reached a height of 10cm. Spacing should be between 50 and 70cm depending on the size and height of the individual variety. Rows should be 75cm apart and the plants should be firmed in after planting to prevent them being rocked in the wind. As the plants grow they will require staking to prevent them being blown over in the wind. Alternatively, some gardeners prefer to mound soil around the base to steady them. Weeds should be often removed from around the plants and watering should be carried out regularly to retain the moisture that they require to produce their good sprouts.

Remove any sprouts that have unfurled, called 'blown' in the horticultural world. Leave the sprouts on the plant until required for cooking as they only keep for a couple of weeks after harvesting. Pick the lower sprouts first and work your way up the plant as the season continues into winter. They can be picked by hand by giving them a sharp, downward tug.

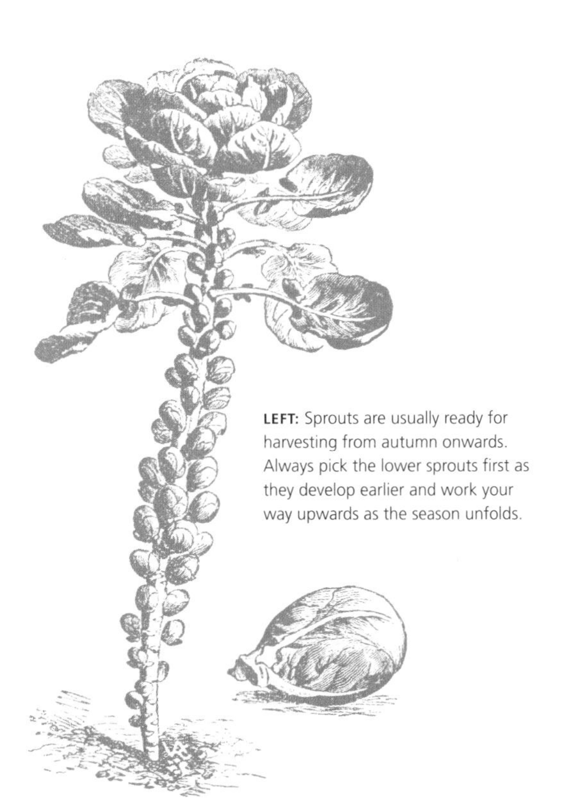

LEFT: Sprouts are usually ready for harvesting from autumn onwards. Always pick the lower sprouts first as they develop earlier and work your way upwards as the season unfolds.

Tasting Notes

Pancetta and thyme Brussels sprouts

The thyme and pancetta in this dish balance out any bitterness of Brussels sprouts. Bacon can be substituted for pancetta if preferred.

Preparation time: 15 minutes
Cooking time: 25 minutes
Serves: 10 people (as a side dish)

- 1kg (2lb) Brussels sprouts
- 3 tbsp olive oil
- 100g (4oz) pancetta, diced
- 1 tbsp thyme, chopped
- Salt and pepper, to taste

Pre-heat a conventional oven to 200°C (400°F / gas mark 6 / fan 180°C).

Blanch the Brussels sprouts in a pan of boiling, salted water for 3 minutes.

Drain and tip into a bowl of iced water to cool quickly. Drain again, quarter and set aside.

Heat the olive oil in a pan. Add the pancetta and cook until slightly crisp.

Slowly spoon in the Brussels sprouts and toss with the oil and pancetta to coat evenly.

Sprinkle over the freshly chopped thyme and season with salt and pepper.

Bake in the oven for about 10–15 minutes until nicely caramelized. Serve.

Kohlrabi

Brassica oleracea Gongylodes Group

Common name: Kohlrabi, stem turnip, turnip cabbage

Type: Annual

Climate: Half-hardy, mild winter

Size: 40cm

Origin: Europe

History: Kohlrabi is a German word meaning cabbage turnip, *kohl* meaning 'cabbage' and *rabi* meaning 'turnip'. Charlemagne, who was crowned emperor of the Holy Roman Empire in AD 800, ordered kohlrabi to be grown in the lands under his reign. Charlemagne, although connected with the French empire, was actually from Aix-la-Chapelle, which is now Aachen located in Germany. This accounts for kohlrabi's German name. Marcus Gavious Apicius, who wrote the oldest known cookbook on cooking and dining in imperial Rome, mentions kohlrabi in his recipes.

The first description of kohlrabi was, however, written by a European botanist in 1554. By the end of the 16th century it was known in Germany, England, Italy, Spain, Tripoli, and the Eastern Mediterranean. It was not until 1734 in Ireland and 1837 in England that kohlrabi was grown on a large scale.

ABOVE: Kohlrabi resembles something more likely found in outer space than in the vegetable garden, but this member of the cabbage family has both an edible swollen stem and leaves.

LEFT: Charlemagne was the emperor of the Holy Roman Empire and in AD 800 he ordered that kohlrabi should be grown throughout his land. The name is German, meaning 'cabbage turnip'.

Cultivation: Sow seeds from early winter onwards in a shallow 1cm drill. Thin the plants out to 20cm spacing once they have germinated. Alternatively, they can be sown indoors in modules and planted out in spring.

Storage: Kohlrabi will keep in the fridge for a couple of weeks, but the leaves should be removed as they leach out the moisture. Ideally, keep kohlrabi outside and harvest as needed.

Preparation: Remove the leaves and top, and tail the swollen stem. Use a potato peeler to remove the skin. It can then be sliced or cut into wedges. To cook, simply boil it in salted water for 20–30 minutes, steam for 30–40 minutes, or fry it in butter. If served raw it can be grated to add sweetness to a winter salad.

This lesser known member of the cabbage family is a gourmet delight for food connoisseurs as it has a distinctive nutty flavour mildly reminiscent of celery. It is hard to find in shops and they can be expensive to buy. The swollen stem can be roasted, steamed or stir-fried and in addition the leaves can also be cooked in the same way as cabbage. It is often cooked in soups but can also be eaten raw in salads and is a good alternative to cabbage in coleslaw. It is a bit of a quirky curiosity with its strange-looking, swollen round stem. The leaves not only sprout from the top but also spiral out from the sides, making a great talking point among gardeners and cooks alike.

It is part of the cabbage family and its mild taste is not too dissimilar to another family member, the turnip, but with a crunchier texture. The skin is usually purple or green, the former usually being slightly hardier, and both have a white flesh.

Like all brassicas, it needs a really fertile, moist but well-drained soil to encourage it to fully develop. They should be grown in full sun and with

LEFT: Purple varieties of kohlrabi, such as 'Purple Danube' can be grown as spring annuals in the flower garden – they are that colourful.

Nutrition

Kohlrabi is low in saturated fat and cholesterol, which makes for a healthier heart and circulatory system. It also contains high levels of vitamins B and C including B6 thiamin, riboflavin, niacin, pantothenic acid and folate, helping to boost immunity, increase metabolic rate and maintain healthy skin and hair.

Tasting notes

Kohlrabi varieties to try

There is quite a bit of variation to try between colour, times of harvesting and flavour.

'Domino' AGM	An early variety that helps to extend the season. Ideal for growing under cover.
'Superschmelz'	A popular cultivar with a mild but sweet-tasting flavour. It can be harvested young and tender or left on the plant to mature to its full size.
'Blusta'	If you are impatient to pick your first kohlrabi then this one is fast maturing and resistant to bolting.
'Kongo'	Quick growing, sweet and juicy and produces high yields. It has an impressive white colour and stays in good condition in the garden more than most other varieties. Good eaten raw.

plenty of well-rotted organic matter dug in well before planting.

Sowing can be started under glass in late winter for an early crop or they can be sown directly in open ground from spring onwards. They usually take about 6–8 weeks to mature. Frequent sowings can be made every few weeks so that there are regular crops from summer until earlier winter.

ABOVE: The edible part of kohlrabi is not a root but a swollen stem that grows just above the ground. The aromatic leaves can be eaten raw or substituted for kale leaves in recipes.

Keep the area surrounding the kohlrabi weed free by hoeing between the rows every one or two weeks. Watering is less important than for other members of the cabbage family, but it is important to keep moisture levels constant as otherwise the plants have a tendency to split.

Vegetables are usually harvested when they are about the size of a cricket ball, but some varieties can get huge, as big as 27kg. Other varieties worth trying include 'Adriana', 'Erko', 'Kolibri', 'Korist', 'Lanro', 'Olivia', 'Quickstar' and 'Rapidstar' or the purple variety 'Purple Danube' as it looks beautiful in the vegetable patch although it is slower growing and can be tougher. If kohlrabi is left for too long in the ground it becomes slightly woody and are much less palatable. Kohlrabi should not need storing as it can be kept in the ground until required. However, like many other roots crops it can be picked and buried in trays of damp sand.

Calabrese and broccoli

Brassica oleracea Italica Group

Common name: Calabrese: green broccoli and broccoli. Broccoli: purple sprouting, purple cauliflower, purple-hearting

Type: Annual or Biennial

Climate: Calabrese is more tender: half-hardy to tender, mild winter. Broccoli is hardier: Hardy, average to cold winter

Size: 45–100cm

Origin: Asia and the Mediterranean region, particularly Italy

History: Broccoli is a member of the cabbage family and evolved from a wilder version in Europe. The history of the wild cabbage dates back over 2,000 years; it would have been a popular food in Roman times. The Italians developed broccoli in the 17th century, and from there it spread across the rest of Europe. Broccoli is still very popular in Italy and the name is derived from *broccolo*, referring to the 'flowering top of a cabbage'. It was the Italian immigrants who brought the plants to America.

Cultivation: Sow calabrese and broccoli from early spring to early summer. They require a fertile, well-drained soil in full sun. Calabrese do not like to have their roots disturbed so should be grown directly in situ 30cm apart. Broccoli can be sown indoor in pots or modules, or in nursery beds, and should eventually be planted out at 60cm apart.

RIGHT: Often called broccoli in the supermarkets, this large-headed green vegetable is properly known as calabrese. It is grown for its delicious flowerheads, stems and leaves.

Storage: They should be harvested before the flowers open and can remain in the fridge for a few days after harvesting. Florets can also be frozen for using later in the year.

Preparation: Remove florets from the plant as required, wash them under running water and then eat either raw in salads, or more commonly boiled or steamed. They can also be added to stir-fries.

You would not be blamed for getting confused about broccoli and calabrese. Calabrese is the large-headed, greenish-blue vegetable with chunky stems that is commonly available in the supermarkets, but confusingly labelled as broccoli. Yet what most gardeners refer to when talking about broccoli is the much less commonly seen purple or white sprouting broccoli that has much smaller florets. As if that is not confusing enough, there is also the quirky looking, lime-green 'Romanesco' variety (see p.84), sometimes referred to as a calabrese, sometimes a cauliflower, and sometimes a broccoli!

Despite the confusion, the good news is that they are all closely related, delicious, easy to grow and are packed full of nutritional goodness, meaning that whichever type of broccoli you grow, you will be rewarded with a vegetable that can be used in a wide range of exciting dishes. They can be added to pasta dishes and coated with garlic and tomato sauces, mixed with chillies and Parmesan. They can also be puréed down into soups. The 'Romanesco' variety has a deliciously rich and nutty flavour and can be made into a version of cauliflower cheese, with lashings of crème fraiche and Parmesan and a topping of bread crumbs. Finally for a touch of indulgence and decadence, try cooking up the florets in batter and dipping them into a sweet and sour sauce.

Nutrition

Like other members of the cabbage family, calabrese and brocolli are some of the most nutritional vegetables you can grow at home. They also have a reputation for their cancer-inhibiting and antioxidant qualities. Along with significant qualities of B complex vitamins, calcium, carotenoids, phosphorus, potassium and vitamin C, they are rich in chromium as well as containing protein and fibre. Recent research indicates that eating sprouted broccoli seeds delivers even greater nutritional benefits.

To retain the health benefits of this wonder plant, it is best to cook it as little as possible, even lightly steaming it for just a few minutes so that it still has plenty of crunch.

Sprouting broccoli

Many people consider the sprouting varieties to have a superior taste to the standard calabrese, which is why it is more commonly seen in public kitchen gardens with restaurants attached, enabling the chef to pop out and harvest florets as and when required. The flavour has been compared by many

BELOW: Purple sprouting broccoli has smaller florets than calabrese and is a useful crop for filling the spring gap in the harvesting calendar, being ready from late winter onwards.

to a mixture of asparagus and cauliflower, and it is stronger flavoured than calabrese. It makes an attractive-looking plant, producing its small edible florets at the top of the plant, but also on side shoots. There are purple and creamy white types, with the former considered to be slightly hardier in the garden. If you are likely to forget to sow seeds each year, then you can try the perennial form called 'Nine Star Perennial', which will keep cropping for five or so years, if it is harvested each season.

ABOVE: Calabrese should be grown in fertile soil in full sun. It is not particularly hardy and so should be harvested before the first of the frosts arrive in autumn.

Sprouting broccoli is wonderfully versatile in the kitchen because the leaves and stalks can also be eaten and added to stir-fries. One of the many benefits of growing this plant is that it provides a crop in the gap between the last of the winter vegetables and before the new season starts.

They have a long growing season, meaning it is probably not an ideal plant if space is short. They take almost a year from when they are sown in mid-spring to harvesting from later winter. However, quick growing catch crops such as radishes and cut-and-come-again leaves can be sown in between rows if space is tight.

Seeds should be sown in spring, either outdoors in nursery beds or in modules in a glasshouse or sunny windowsill. If sowing in pots, place two seeds in each and thin them out to one after germination. In a nursery bed they should be sown thinly in shallow drills and thinned out to 15cm after germination. When the seedlings are about 8 to 10cm tall they can be planted out in rows at their final spacing of 60cm between each plant, and 60cm between each row. Plants that were raised indoors will benefit from being hardened off for a few days in a cold frame or porch before being planted outside.

Purple sprouting broccoli requires a sheltered position in full sun. Like other members of the cabbage family, they are not ideal in acidic soil due to their susceptibility to the club root fungus. If this is a problem, then a dressing of lime should be applied to increase the pH to about 6.5 to 7. The plants will need to be kept weed-free to avoid competition for nutrients and moisture.

Harvesting begins in late winter. The tiny florets should simply be snapped off as they begin to swell, but before they open up as flowers. If the flowers do open, the plants at least look attractive with their rich yellow flowers against the dark green foliage early in the year. Regularly pick over the plant from late winter to mid-spring, because if the buds do open the plant will go over quickly and stop producing a crop. Excess amounts of florets can be frozen.

Tasting notes

'Romanesco'

The 'Romanesco' is a bizarre-looking variety, with tightly whorled, pointed florets in an almost alien-type of florescent green. The texture looks like something you are more likely to see as coral at the bottom of an ocean bed, rather than in a vegetable bed. Among connoisseurs it is supposed to have the finest flavours of all the broccoli/cauliflower types, with a sweet herbaceous nuttyness. In many potagers and kitchen gardens it is used almost as much as a decorative feature as it is as an edible crop. It is hardier than calabrese and is usually harvested in the gap after calabrese has finished in autumn and before the sprouting types have started in late winter.

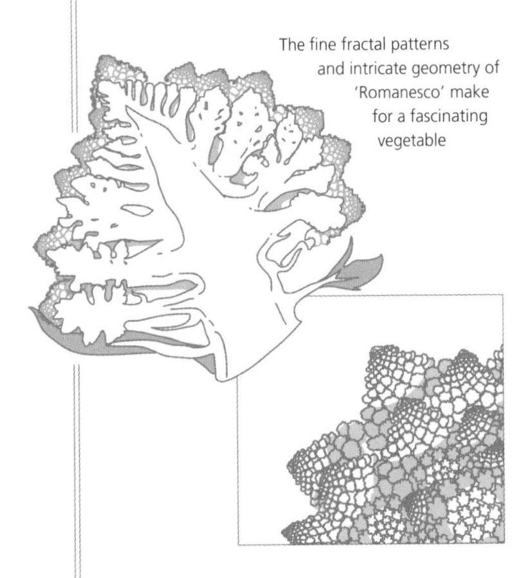

The fine fractal patterns and intricate geometry of 'Romanesco' make for a fascinating vegetable

ABOVE: 'Romanesco' a fantastic but quirky addition to the kitchen garden and is a useful 'filler' as it is ready for harvesting after calabrese and before purple sprouting broccoli.

How to grow calabrese

If you are short of space, yet want to grow a broccoli-type of vegetable, then grow this one rather than the purple/white sprouting type. It is much faster growing and can be harvested in time for autumn crops to be sown and planted, such as broad beans and garlic. However, it is not particularly hardy, and must be harvested before autumn. It is grown for its one large central head, although after harvesting some varieties will continue to produce side shoots with smaller florets. 'Romanesco' types (see box) do not produce side shoots after being cut and should be dug up and added to the compost heap after harvesting.

Calabrese requires a fertile site in full sun with a well-drained soil. Add lots of organic matter to the soil well before planting as this will help to retain the moisture during summer. Unlike brocolli, calabrese do not like to have their root system disturbed and for that reason are sown directly in the place they are to grow for the season.

Use the corner of a draw hoe to create a shallow drill and then sow 'stations' or 'clusters' of three seeds every 30cm. Rows should also be 30cm apart. After germination, the seedlings should be thinned out to 1 per station. Calabrese should be kept weed-free and well watered during the summer.

The heads should be ready for harvesting about 12 weeks after sowing and be cut before the buds open. Keep an eye out for subsequent smaller heads from the side shoots that can also be harvested and eaten. The calabrese should be eaten soon after harvesting, but if there is a glut, the head can be chopped up into florets and frozen.

Mizuna

Brassica rapa var. *nipposinica*

Common names: Mizuna green, shui cai, kyona, Japanese mustard, potherb mustard, Japanese greens

Type: Annual

Climate: Half-hardy, mild winter

Size: 30cm

Origin: China

History: Mizuna is native to China, though it is considered a Japanese green as it has been cultivated there for several centuries. The name for this leafy green comes from the Japanese *mizu*, meaning 'water', and *nu*, meaning 'mustard plant'.

BELOW: Mizuna hails from China and is popular in Asian cuisine. It has a slightly mustardy flavour and can be grown as a cut-and-come-again plant, harvesting leaves regularly as required.

Nutrition

Mizuna greens are an excellent source of vitamins A, C and E (in the form of carotenoids) and manganese. They aid in the prevention of cancers due to their anti-inflammatory, anti-oxidant and detoxifying properties.

Cultivation: Sow directly into the soil between March and August in shallow drills about 1cm deep. Grow as cut-and-come-again leaves and thin seedlings to 5cm apart, or grow as larger plants and thin to 20cm apart.

Storage: Like any salad leaf, mizuna does not last long after being put in the fridge so are better being harvested as and when they are needed in the kitchen.

Preparation: Mizuna has a mild flavour that is almost sweet, with a faint hint of a mustardy tang. When fresh and in good condition, the plant is crisp with a bright, clear flavour and a hint of a crunch, adding texture as well as flavour to the dishes it is included in. It is an Asian green that can be eaten raw (in salads) or cooked.

Mizuna and mibuna are two types of 'cut-and-come-again' Japanese leafy vegetables that are becoming increasingly popular with foodies who like to pep up their salads or stir-fries with peppery, mustard-flavoured leaves. Very similar to mizuna, but with more of a punch, is the similarly named mibuna. Both are worth trying in raw in salads or side dishes, or adding to soups.

Mizuna and mibuna plants can either be grazed as cut-and-come-again every few weeks after sowing, or if left in the ground they form heads or rosettes with the dissected leaves. Looking very much like rocket, they can be harvested with a pair of scissors at the base of each plant when young, and allowed to re-sprout. Or they can be left to mature to form semi-mature or mature cabbage heads. The foliage is decorative with its serrated, dark leaves held aloft on white stalks, making them both a popular feature for edging flower borders and garnishing dinner plates. A variety called 'Mizuna Purple' has particularly attractive purple stems. The young flower stems can also be cooked.

Whereas pak choi can be prone to bolting in dry weather, mizuna and mibuna tend to be less prone to this problem. Mizuna and mibuna are easy to grow and tolerate moderately cool and damp conditions, but they can struggle in very dry conditions and will need frequent watering every few days.

Seed should be sown directly into the soil any time between March and August in shallow drills about 1cm deep. Rows should be 30cm apart. It is best to make regular sowings throughout the season so that the salad leaves are ready to harvest at different stages throughout spring and summer. To do this, grow them in short rows every 2–3 weeks, rather than just one sowing in a single long row. Remember to eat the thinnings as micro greens – see p.85 for how best to space the plant.

For a really early crop it is possible to grow them under glass in February in pots. Place them in a cold frame to harden them off for a few days before planting them outdoors.

Cut-and-come-again crops can usually be harvested up to fives times for baby leaves throughout the season, after which time they can be dug up and composted or left to grow large. Larger heads should be harvested just below the rosette, using a sharp knife.

Bolting

Bolting is often caused by the plant becoming stressed, often by a hot or cold spell. Usually it is caused by a dry spell just after sowing the seed. It causes the plant to panic, send out a flower shoot and go to seed very quickly. Usually the few leaves that have been produced taste bitter and are not worth eating. Sowing early when it is cooler can help to mitigate this problem. Alternatively there are bolt resistant varieties that can be grown. Finally, regularly watering the plant after sowing should help to avoid the problem.

Canna lily

Canna indica

Common name: Canna lily, Queensland arrowroot, Indian shot plant

Type: Rhizomatous perennial

Climate: Half-hardy, mild winter

Size: 1.6m

Origin: North and South America

History: Canna lilies have been used as a food crop for thousands of years, but they were not well known to European botanists until the 1500s, and it was not until the late 1800s, the Victorian era, that they became widely popular as ornamental plants. Unfortunately, these lilies were largely lost in gardens because European gardeners stopped growing cannas during the upheaval from World War One through World War Two. In addition, garden fashions changed. In the first half of the 20th-century garden designers, such as Gertrude Jekyll, replaced formal-looking Victorian gardens with informal, relaxed perennial borders. This led gardeners to abandon many plants used by the previous generation, including the canna.

Cultivation: Grow them in fertile, well-drained soil with lots of well-rotted organic matter. Grow them from rhizomes or as plants bought from a nursery or garden centre.

Storage: Store the rhizomes in a dark place such as a paper bag in a cupboard for a few weeks until ready to use.

Preparation: Canna root does not store well so it is best left in the ground until ready for use. It can be eaten raw, but is often boiled. The best method of preparation is to bake it. Rinse the canna root in clean water, dry it using a paper towel, then cut it into small pieces. The root may be cubed, diced or julienned. Fill a pot with water and bring to the boil, then place the root into boiling water for 10–15 minutes. Cooking time will vary based on how small the root is chopped.

RIGHT: The beautiful ornamental canna lily has deliciously tasty roots. Do not confuse it with other types of lilies that are not edible.

Cannas are more commonly seen in subtropical borders and bedding displays than in the kitchen garden. Their huge, lush and glossy leaves hold aloft spikes of fiery coloured flowers, yet underground, cannas offer a gourmet alternative for people who like to try something a bit different. This Amazonian root vegetable produces sweet rhizomes that can be cooked and treated like the humble spud or a Jerusalem artichoke. Whether you prefer mash or French fries, simply substitute the roots of this exotic-looking plant for the potato, although they usually need boiling for longer to break down the more fibrous structure. Canna crisps (see box) are great served to impress friends. In South America the leaves are wrapped around meat, fish and poultry; they are also used in rice dishes and can be baked, steamed or grilled.

Tasting notes

Canna crisps

A versatile snack for all occasions. Once baked, you can flavour the crisps to suit – why not try a dusting of hot and spicy paprika?

Preparation time: 10 minutes plus 2 hours of soaking
Cooking time: 15 minutes
Serves: 2 people (as a snack)

- 1kg (2lb) canna roots
- 2 tsp salt
- 2 tbsp oil

Clean the canna roots and peel with a potato peeler. Then slice them into paper thin slices, either using a mandolin or sharp knife.

Place the slices, canna roots in a large bowl of salted water and leave to soak for a few hours to remove some of the excess starch.

Strain the slices. Leave them to dry thoroughly.

Pre-heat a conventional oven to 120°C (250°F / gas mark 1 / fan 100°C).

Heat up oil in a wok or frying pan over high heat and start to fry up the chips until they turn a golden colour.

Place them on a baking tray lined with baking paper and cook them in the oven for 10 minutes.

LEFT: Canna lilies are beautiful, architectural plants with bright flowers. Their roots are commonly known as arrowroot.

Plant cannas in a sunny, sheltered site in well-drained soil. Several weeks before planting, dig in lots of organic matter, such as well-rotted manure into the ground, as the large glossy leaves need plenty of moisture and nourishment to keep them going.

For those on a lower budget, the plants can simply be grown from a packet of rhizomes rather than buying plants. In March, place individual rhizomes into plastic pots full of general-purpose compost and keep them in a warm place in a heated glasshouse, conservatory or windowsill, remembering to water them regularly. They can be planted out in late spring after the risk of frost has passed. However, if purchased from the garden centre, avoid eating them for the first year as they may have been treated with chemicals. Wait until the following year, when the chemical should no longer be present. There are lots of different varieties to try and some will be better than others, so it is worth experimenting with a few different types.

Some of the taller plants will need to be staked to prevent the plants collapsing on each other. Keep them weed free throughout the growing season and regularly watered.

ABOVE: Harvesting the canna lilies' roots should be done towards the end of the growing season when the plants are tall and mature.

Perpetuating cannas

- Gently dig up canna rhizomes with a fork in the autumn months, taking care not to damage the root system.
- Select some of the plumper rhizomes for cooking by carefully breaking them off – do not remove more than about a third.
- Leave the remaining rhizomes attached to the plant – these are needed for next year.
- Cut back the stems and foliage.
- Place the remaining rhizomes in compost or moist sand over winter in a cool, frost-free place. Do not let them get too wet as they will rot, but at the same time do not let them dry out.
- Plant the overwintered rhizomes out again in late spring.

In milder areas overwintering cannas do not have to be stored indoors but can simply be replanted after harvesting. For frost protection, cover them with a 15cm-deep mulch to protect the base of the plant during winter. There may be some losses but it is much easier than having to store them over winter.

SMALL SPACES

Vegetable growing is possible in the tiniest of spaces. If space is really restricted, then it makes sense to only grow the vegetables that you really love and cannot buy in the shops. In many ways, growing in a small space is better as it helps concentrate your efforts on the vegetables you really want to grow, it avoids gluts and, for people who are time-poor maintenance and management is far easier. Vegetables can be grown anywhere including roof gardens, balconies and small courtyard gardens.

GROWING IN SHADE

In many urban gardens, shade is a problem as there are lots of buildings casting their shadows over all or part of the garden. This can initially make it appear tricky for vegetable growing, but there are many crops that will tolerate shade. They generally tend to be the leafy crops such as lettuces, spinach, Swiss chard and cut-and-come-again crops, as they produce more leaves when light levels are lower. Plants that produce fruits such as tomatoes, aubergines, squashes and courgettes should be avoided. The cabbage family such as sprouts, broccoli, kohlrabi and kale will only tolerate light shade. Most of the root family including carrots, beetroot, turnips and potatoes require at least half a day of sun.

Rhubarb thrives in shady conditions, needs hardly any attention, and will smother out any weeds.

ABOVE: Rhubarb will tolerate shade, making it ideal for growing at the base of north-facing walls and fences, or in the shade of a shed or house.

VERTICAL WALLS

Structures can be attached to walls with irrigation systems, enabling vegetable plants to be grown on vertical structures. It is important to ensure that the vegetables are either shade tolerant or are not casting shade over each other. Careful monitoring of their watering requirements is also needed.

CONTAINERS

Vegetables can be grown in almost any container, so long as there is enough space for the roots to develop and they have a drainage hole in the bottom. Vegetables grown in containers will require much more watering and feeding than if they were grown in the ground. In warm weather they may need watering as often

LEFT: Vegetables can be grown in almost any type of container. Even an old gardening boot with a drainage hole is suitable.

'The possessor of an acre, or a smaller portion, may receive a real pleasure, from observing the progress of vegetation ... A very limited tract, properly attended to, will furnish ample employment for an individual.'

Vicesimus Knox, *Essays Moral and Literary*, (1778)

as once or twice a day and feeding once a week with a liquid fertilizer during the growing season. One of the benefits of growing in a container is that they can be moved into the shade if the heat gets too much. Containers may also benefit from being turned during the day so that all sides of the container receive the sun.

RAISED BEDS

If your back or front garden is covered in concrete or patio slabs, then do not despair. Raised beds could be the answer. Growing vegetables in this manner is low maintenance, allows for easy weeding and saves on all that back-bending work. Vegetables in raised beds usually have better drainage and ripen earlier as the soil within them warms up more quickly. The raised beds should be filled with the very best, loam-based compost, meaning the vegetables have the greatest possible growing conditions.

USING RECYCLED MATERIALS

Potatoes can be grown in an old dustbin or in a stack of old car tyres, with more types and soil simply added to the stack as the foliage grows. They can also be grown in large builder bags, simply unrolling the bag and topping up with more compost as they grow.

Plants such as courgettes, pumpkins and squashes can also be grown on the top of builders bags filled with compost.

Plants suitable for window boxes

A window box just outside the kitchen is ideal for growing leaves. They are easy to maintain and regular sowings can be made every few weeks to ensure there is always a plentiful supply.

Most vegetables can be grown in containers, but some are better than others for window boxes as they are more compact and require a shallow root area. Tall plants will block the view from your window.

Here are some of the best vegetables to grow in a window box:

Lettuce, radish, beetroot, rocket, mizuna, spring onion, chives, spinach, carrot (such as globe types or dwarf chantenay types), watercress in damp soil and trailing tomatoes.

BELOW: Pumpkins and squashes are hungry feeders and have traditionally been grown directly on compost heaps, an ideal space- saving solution.

Pepper and chilli

Capsicum annuum Longum and Grossum Groups

Common name: Pepper and chilli, chilli pepper, capsicum

Type: Annual

Climate: Tender, warm-temperate glasshouse

Size: Between 25cm–1m

Origin: South and Central America

History: Chilli peppers were perhaps one of the first plants to be domesticated in Central America, where there is evidence that they were consumed in 7500 BC. They were introduced to South Asia in the 1500s and have come to dominate the world spice trade. India is now the largest producer of chillies in the world.

Cultivation: They should be sown indoors in late winter or early spring and then grown under glass in cold areas. Growing tips should be pinched out when the plants reach about 20cm to encourage a bushier plant. Outdoors they can be planted directly into fertile, free-draining soil or in grow bags or containers in a warm, sunny location in more favourable climates.

LEFT: Chillies require a long season for them to fully develop their spicy flavours, so seeds should be sown early on in the year under cover.

RIGHT: Peppers have a wonderful ornamental quality and develop into an array of different colours, including green, yellow, orange, red and purple.

Storage: Peppers unfortunately go mushy if frozen, although chillies tend to fare better. Peppers will keep for a couple of weeks in the fridge. The best way to preserve chillies is to dry them out in the sun on a wire mesh, such as chicken wire, or to hang them from strings and allow them to dry. Another alternative is to use them to infuse cooking oil.

Preparation: Chilli peppers contain oils that can burn your skin and especially your eyes, so it is important to be very careful when handling them. It is a good idea to wear gloves when preparing hot chillies and, whatever you do, do not rub your eyes. Slice chillies in half lengthways and remove the seeds before chopping them finely. Cut out the core from peppers and slice or dice.

Tasting notes

Sticky chilli jam

This chilli jam is the perfect accompaniment to cheese and crackers or can simply be spread on crusty French bread.

Preparation time: 10 minutes
Cooking time: 1 hour
Serves: makes 3–4 ½l (1lb) jars

- 400g (13oz) cherry tomatoes
- 9 red peppers
- 10 red chillies
- 7 garlic cloves, peeled
- 4cm (1½in or thumb length) of root ginger, peeled and chopped
- 750g (1½lb) sugar
- 250ml (8fl oz) red wine vinegar

Place the tomatoes, peppers, chillies, garlic and ginger into a food processor and whizz until finely chopped.

In a pan, dissolve the sugar in the vinegar over a low heat.

Add the tomatoes, peppers, chillies, garlic and ginger mix and simmer for about 40 minutes, or until the liquid has reduced and it has a thick, sticky consistency.

Once the jam is becoming sticky, cook for 10–15 mins more, stirring frequently.

Cool slightly, then transfer into sterilized jars.

Once re-opened, it will keep for about 1 month in the fridge.

ABOVE: Peppers should be grown indoors in cooler areas of the country. In a warm and sheltered spot they can be grown outside in full sun.

Some like them hot and spicy, others like them sweet and crunchy; whatever your taste there is a chilli or pepper for everyone. The two types are very closely related; peppers are milder and larger while chillies are usually hotter, although there are varieties that are gentler on the taste buds. The popularity of chillies has grown thanks to the increase in popularity of Indian, Thai, Chinese and Mexican dishes over the last few decades. Mediterranean food is also often flavoured with both peppers and chillies as well as the spicy paprika powder that is extracted from this plant once it has dried out. It is mainly the pith that provides the knock-out fiery punch, so wash and remove the seeds if you want something milder.

Peppers and chillies require a warm, sunny position outdoors in mild areas. In cooler regions they may have to be grown in an unheated polythene tunnel, glasshouse or conservatory.

They need a well-drained but moist soil, that should ideally be slightly acidic. Lots of organic matter should be added to the soil as this helps to retain the moisture. Most people however, do not grow these plant directly in the ground, but instead grown them in containers filled with general-purpose compost or growing bags. If using the latter, then no more than two plants should be planted per bag.

Sowing takes place indoors in pots in a heated propagator or a warm and sunny windowsill. Chillies need a longer growing season to achieve their heat so should be sown in late winter. Peppers can be sown a few weeks later. They should be transplanted into 9cm pots when they have produced their first two true leaves. Once the risk of frost is over, they should be hardened off in a cold frame for a few days before being planted out at 45cm apart. When the growing tips reach about 20cm they should be pinched out to encourage a bushy plant, which in turn will produce a larger crop. Plants will require regular watering, although avoid over-watering chillies too close to harvest time as it can dilute the heat.

You can expect to get up to 5–10 peppers per plant, whereas chillies will produce a few dozen depending on variety and growing conditions. Fruits should be harvested when they are green to encourage the plant to produce more. They will change colour if left on the plant, with peppers turning red, yellow, orange and purple and becoming sweeter, while chillies will become hotter.

They are usually ready for harvesting outdoors from August and will continue to crop in a favourable location until autumn.

The Scoville heat scale

The Scoville heat scale measures the compound called capsaicin, that gives chillies their heat. The hotter the chilli, the higher it scores in the scale. It was developed by Wilbur Scoville in 1912. Prior to this scale, the heat of chillies was simply done by taste.

The Guinness World Record holder for the hottest chilli is currently Smokin Ed's 'Carolina Reaper', grown by The PuckerButt Pepper Company (USA), which rates at an average of 1,569,300 Scoville Heat Units (SHU).

As an idea of scale, a Scotch Bonnet scores between 100,000 and 300,000; a Tabasco pepper between 30,00 and 50,000; a Hungarian wax pepper 3,500 to 8,000; a pimento between 100 and 900; and a bell pepper scores 0.

BELOW: Chilli peppers were one of the first crops to be cultivated in Central America. Chillies get hotter as they mature. Peppers get sweeter the longer they stay unpicked.

Chop suey greens

Chrysanthemum coronarium

Common name: Chop suey greens, chrysanthemum greens, garland chrysanthemum, crown daisy, kikuna, mirabeles and shungiku

Type: Annual

Climate: Half-hardy, mild winter

Size: 25cm

Origin: Mediterranean, East Asia

History: In the 8th century, the chrysanthemum was introduced into Japan where it was quickly named as the national symbol of Japan by the emperor and became the inspiration for the royal seal. Carolus Linnaeus named it by taking the Greek word *chrys*, which means 'gold coloured', and adding *anthemon*, which means 'flower'.

Cultivation: Sow seeds directly in the soil in shallow drills in mid-spring. The site should be in full sun and the soil should be free-draining and fertile. Harvest leaves like a cut-and-come-again plant, and make regular sowings throughout the season so that there is plenty to harvest.

Storage: Leaves will only last for a day or two after picking so only harvest on days when they are going to be used in the kitchen.

Preparation: Simply wash and cut the leaves or eat them whole, raw or cooked.

This is a fast-growing and attractive form of the garden chrysanthemum and is grown for its delicious, dissected young leaves, which can be added to salads. There is a double-whammy in that the flowers can also be eaten, but if you let them flower then the flavour in the leaves turns bitter and unpleasant. They have a mildly herbal and nutty flavour and are eaten raw or lightly cooked. The leaves are best picked young when they can be steamed or added to stir-fries. The leaves are

popular in Japanese soups but their attractive shape and colour makes them a great visual feature in the salad bowl or as garnish too. The flowers, both fresh and dried, and the unopened buds can also be used to make an aromatic herbal tea.

With deeply serrated leaves and attractive pale yellow and orange chrysanthemum flowers, this is an attractive plant for vegetable gardens and potagers, but is just as at home in the flowerbed. It is a low-growing plant and so is a good choice for separating out areas of the kitchen garden and for edging paths. It can be interplanted with other larger crops, such as sweetcorn, among other cut flowers, and also in brassica beds where it contrasts well with the textures of cabbages and kale. The light green foliage also looks good when grown next to red cabbages or red chicory. This must be one of the easiest vegetable plants to grow and the leaves can literally be harvested a mere six weeks after sowing.

If you cannot find the seeds in the vegetable seed catalogues, then try the ornamental and flowering sections as they are very often listed there. Sow them outdoors from mid-spring onwards in fertile, well-drained soil in full sun, 1cm apart in shallow drills, with 30cm between the rows. Thin the seeds out to 20cm apart when they are large enough to handle. Seeds should be sown regularly every couple of weeks to ensure a regular crop of chop suey greens, but avoid midsummer as the dry weather can cause the plants to bolt and prematurely flower.

LEFT: Chop suey greens are a type of chrysanthemum and are just as likely to be found in an ornamental flower bed as they are to be grown in the kitchen garden.

To get an early crop, seed can be scattered indoors in trays of seed compost at 0.5cm deep, before hardening them off in cold frame for a few days prior to planting them out at 20cm apart outdoors.

A few weeks after sowing the chop suey leaves will be ready for harvesting. They can be harvested like a cut-and-come-again plant by cutting leaves with scissors above the base of the plant and allowing it to resprout for harvesting a few weeks later. If you have a number of plants, leave some uncut so that their flowers can be enjoyed later to add to salads or for making tea.

Endive
Cichorium endivia

Common name: Endive

Type: Annual

Climate: Half-hardy, mild winter

Size: 30cm

Origin: Middle East

History: It is thought that endives originated somewhere in the Near East and the ancient Egyptians were thought to have cultivated them. In ancient times, endive and chicory were classed as the same plant and that they have always been eaten as a salad crop. They became a popular crop in southern Europe before spreading into northern Europe.

Cultivation: Endives should be sown in spring, in shallow drills, although the broad-leaved types are more winter hardy and can be sown in late summer for a winter harvest. Cloches should be placed over the seedlings once the autumn frosts arrive. The curly-leaved types are often blanched before harvest by placing a pot over the plant. This reduces some of the bitterness of the leaves (see p.100).

Storage: The leaves only last for a few days in the fridge once picked, so it is best to harvest little and often as required in the kitchen.

Preparation: Chop up the leaves and add raw to salads or leave whole and boil by plunging into salted boiling water for about 5 minutes.

Nutrition

Among all the green vegetables, endive is one of the richest sources of vitamins A and C. It also contains calcium, chlorine, iron, phosphorus, potassium and sulphur and is an excellent provider of fibre and carotene.

Endive is a popular leaf vegetable that can be eaten either raw, although it can taste slightly bitter, or sautéed, added to stir-fries, soups and stews. There is often a huge amount of confusion between the two members of the daisy family, chicory and endive, but that is not really surprising. After all, in France and America endive is what the

BELOW: Believed to have originated somewhere in the Near East. The ancient Egyptians are known to have grown them, but their popularity spread to Southern Europe, and then further north.

English call chicory and very often the name is used interchangeably in restaurants to describe bitter or blanched leaves. Even more confusingly, the vegetable that is sometimes called Belgium endive is a form of chicory. However, there is one important distinction to the gardener; endive is an annual that needs to be sown each year, whereas chicory is a perennial that will provide a crop year after year.

TASTING NOTES

Endive and prawn omelette

Endives are low in calories and provide a great chance to include less-known vegetables in a healthy diet. Try combining with eggs to make an alternative to the traditional omelette.

Preparation time: 5 minutes
Cooking time: 10 minutes
Serves: 1 person

- 3 large eggs
- 15g (½oz) butter
- 2 endives, chopped
- 100g (4oz) prawns, cooked and peeled
- Salt and pepper, to taste

Beat the eggs in a bowl. Add pepper and salt.

Melt the butter in a frying pan and fry the endives; season with pepper and salt.

Pour the beaten eggs on to the endives and cook on low heat. Make sure not to overcook.

Sprinkle the prawns on top and serve.

RIGHT: Escarole or bativan endive has broad, deep green leaves and are hardier than the frizzy types, making them suitable for growing during winter.

There are two distinct types of endive available to the gardener. Firstly there is the frizzy-leaved type, which has attractive frilly, tightly-curled leaves. It is often called *frisée*, or just to confuse matters more, the French call it *chicorée frisée*, but just to clarify, it is an endive. It looks stunning when grown in a potager or even edging a flower bed, with the foliage creating a wonderful textured effect. However, it is not winter hardy, so it is grown just as a summer crop. The other type is sometimes called Batavian endive or escarole which has broader, deep green leaves that are slightly wavy. These are much hardier than the crinkly type and are suitable for growing during winter. Endive is either grown as a cut-and-come-again plant, where two or three harvests can be made throughout the the year, or it is harvested when it has formed its rosette of leaves.

This popular leafy vegetable does have a slightly bitter taste and this is far more noticeable in a hot summer. It is not so commonly grown as lettuce, one of the main reasons being that it lacks the sweetness. However, the one benefit it has over lettuce is that it can be grown as a winter crop, whereas lettuce can sometimes struggle. Also, the slight bitterness does appeal to some people and provides a nice contrast to some of the other sweeter leaves in mixed salads. Endives are also generally less prone to diseases or bolting.

Endives should be grown in fertile, well-drained soil in full sun. The site should have good moisture retention, however, so that an abundance of leaves is produced. Dig over the soil prior to sowing seeds and add plenty of well-rotted organic matter. Both types of endives, flat and frizzy, should be sown in spring, although for a winter crop the flat type can be sown in late summer. Seeds should be sown in shallow drills that are about 1cm deep, with rows 40cm apart. Once the seedlings appear they should be thinned to approximately 20cm apart for the frizzy types and 40cm for the larger broad-leaved endive. The plants must be kept well watered during dry periods or they could either wilt or bolt. They will also benefit from a liquid feed every couple of weeks. Regularly hoe between the rows to prevent the developing plants from being smothered and diminished by competing weeds. The winter varieties should be covered with a tunnel cloche before the arrival of the autumn frosts.

The leaves can be harvested like a cut-and-come-again plant, by snipping off a few leaves as and when needed. Alternatively the entire plant can be removed and taken into the kitchen.

Blanching the curly-leaved varieties

Blanching the leaves can reduce the amount of bitterness they contain. It is done by placing a plastic pot, bucket or terracotta pot over the plant for a week or two before harvest during summer. If using and old plant pot, make sure the drainage holes are covered up too. Sometimes simply placing a dinner plate, kitchen tile or even a piece of cardboard over the leaves is enough to blanch them, but make sure the leaves are dry first because otherwise they can rot and quickly attract slugs and snails. Plants are usually ready for blanching about 12 weeks after sowing, once the heads have started to form. The broad-leaved type of endive can also be blanched in the same way, but because of the thickness of the leaves it is often just as effective to tie up the leaves into bunches, which prevents the sunlight from reaching the central leaves and therefore blanching them.

LEFT: Endives produce stunning, decorative flowers. These will appear on the plants if they are left to grow, and are very distinctive of the genus *Cichorium*.

Chicory

Cichorium intybus

Common name: Chicory, radicchio, chicons, Belgium chicory, witloof, sugar loaf chicory

Type: Perennial

Climate: Hardy, average winter

Size: 25cm

Origin: Northern Africa, Western Asia, Europe

History: Chicory is one of the oldest recorded types of plants. Its cultivation is thought to have originated in Egypt in ancient times, later being grown by Medieval monks in Europe. In Europe during the 1820s the roots of chicory were baked and ground then used as a substitute for expensive coffee beans. They are still used as an additive to coffee today. Some brewers also used roasted chicory, adding it to their stout beers for extra flavour.

Cultivation: Seeds are sown in spring in shallow drills in a sunny, sheltered aspect in well-drained soil, with harvesting taking place in late autumn or winter. To force the tender chicons during winter, a few plants should be dug up, taken inside, placed in a bucket and covered with compost to be harvested three weeks later.

Storage: Chicory does not last for long in the fridge, but its shelf life can be extended by placing it in a paper bag to exclude light as this prevents the chicory discolouring and turning even more bitter. Ideally the heads should be harvested as needed from the garden.

Preparation: Chop into pieces or leave whole. To cook, simply plunge the heads into salted boiling water for about 5 minutes.

RIGHT: Belgium chicory is also known as witloof chicory and is grown for its crunchy green leaves. The plants is also used for forcing to make chicons.

Chicory is particularly popular in Italy and was enjoyed by the Romans. It is closely related to endive and very often the two are confused. Both are grown for their bitter salad leaves and both supply useful winter leaves for the kitchen. However, the one main distinction is that the endive is an annual whereas chicory is a perennial, although it is often grown as an annual in the vegetable garden. Chicory is also an attractive addition to the vegetable garden, particularly the radicchio types, and when allowed to flower provides a beautiful display of bright blue flowers. These should be removed before they seed otherwise the plants will seed themselves freely around the garden.

ABOVE: If allowed to flower, chicory produces the most exquisite lavender-coloured flowers. The roots of the plant are often dried and used to make a caffeine-free version of coffee.

BELOW: The succulent leaves of this herbaceous perennial can be added to winter salads. Chicory also produces a large tap root below the ground, similar to a dandelion, which persists after the leaves are harvested.

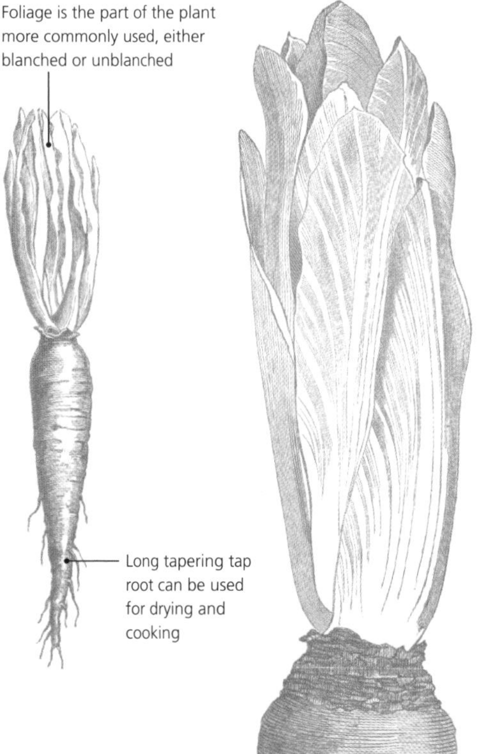

It is thanks to its perennial root system that it is also grown as a coffee substitute or mixed with coffee to bulk it up. Like dandelion, it can be dug up, and its root removed from the plant, washed and roasted and made into a thirst-quenching hot drink. However, most gardeners grow chicory for the leaves. There are three distinct groups: Belgium or witloof chicory, sugarloaf and radicchio. Witloof chicory produces edible green leaves, but they are more commonly grown as the gourmet treat known as chicons (see box). Popular varieties include 'Brussels Witloof' and 'Witloof Zoom'.

Forcing chicons

Witloof or Belgium chicory can be forced indoors to create delicious tender, white chicon heads. To do this the leaves should be cut back hard in late autumn, and the roots gently dug up out of the ground and taken indoors where they should be planted in the bottom of a bucket, container or even a compost bag. They should then be covered with compost so that just the growing tips show, and then covered with another bucket or container and left in the dark for about three weeks. The emerging chicons can be harvested and the remainder of the plant can be covered again in order to repeat the process. Chicons should be harvested when they are about 15cm long. The popular variety for forcing chicons and blanching is 'Brussels Witloof'.

Chicory is harvested from late summer and into autumn. Covering them up with cloches will extend the season.

Chicons are used in many gourmet dishes, but *Chicons au gratin* is particularly popular in Belgium and is best served with lots of strong-flavoured cheese and a béchamel sauce. If you cannot get hold of chicons, then leeks make an equally suitable alternative.

The sugarloaf chicories produce a rosette of broad leaves, looking similar to cos lettuce with its large upright head. It is a useful crop as it is usually harvested in autumn when many salad leaves are not available, and if it is protected with a cloche will continue well into winter. Their upright habit means that the inner leaves are self-blanched by the outer ones, making them much sweeter. Reliable, tried-and-tested varieties worth trying include 'Pan di Zucchero' and 'Zuckerhut'.

The radicchio types are probably the most ornamental of all the chicories and are the ones often seen in mixed bags of salad from the supermarket. Sometimes called red-leaved chicory, it adds a beautiful splash of colour to the kitchen garden. Radicchios also brighten up winter salads as they grow well under cloches, where they can survive well into winter. 'Indigo' is a popular variety producing a dense head with red hearts and dark green outer leaves. 'Palla Rossa' produces large heads and has impressive red hearts.

LEFT AND ABOVE: Red-leaved chicory is also known as radicchio, and adds a splash of colour to the kitchen garden. Usually harvested in autumn, the season can be extended further with cloches.

All the chicories require a fertile soil in full sun. Plenty of organic matter should be dug into the soil before sowing. Witloof chicory is sown in late spring or early summer into shallow, 1cm-deep drills 30cm apart. After germination the seedlings should be thinned out to 15cm and any emerging flower stems should be removed. To blanch the leaves outdoors, the leaves should be cut back in early autumn and the remaining crown should be covered in a 15cm deep mulch. The young leaves should be picked as they emerge from the ground.

BELOW: Chicory is a useful crop in the garden as it has attractive flowers, edible leaves for salads and roots that can be dried to create an alternative hot drink to coffee.

Tasting notes

Grilled chicory with pears and hazelnuts

The sweetness of pear and the slight bitterness of chicory make them the perfect accompaniment to each other, while the hazelnuts put the crunch into this salad – an ideal starter for a casserole or stew.

Preparation time: 10 minutes
Cooking time: 10–15 minutes
Serves: 4 people

- 2 large heads of chicory
- 2 tbsp olive oil
- 1 ripe pear
- 2 tbsp hazelnut oil
- 1½ tsp thyme, chopped
- Pepper, to taste
- 25g (1oz) hazelnuts
- 1 tsp thyme sprigs

Cut the chicory lengthways and remove core.

Brush with olive oil and place in a grill pan with cut side up. Grill for about 3–4 minutes. Turn and baste. Cook for 2–3 more minutes.

Halve, core and slice the pear.

Turn the chicory and top with the pear slices.

Brush with hazelnut oil, add the thyme and pepper, and grill for 5–6 minutes.

Garnish with hazelnuts and sprigs of thyme and drizzle with the remaining hazelnut oil.

Sea kale

Crambe maritima

Common name: Sea kale, *Crambe*, sea-colewort, scurvy grass, halmyrides

Type: Perennial

Climate: Hardy, cold winter

Size: 50cm

Origin: North and west coast of Europe

History: A wild plant that grows along many of the shorelines of Europe. Sea kale has been around for quite a time, the taste of which resembles cabbage. Louis XIV knew of it, and ordered its cultivation in the gardens of Versailles. It has been cultivated in the gardens of England since the 18th century and was mentioned in Thomas Jefferson's *Garden Book* of 1809. It was a popular food, often served at the tables of the very rich, until the early 20th century, but today it is no longer as popular.

Cultivation: Stems should be forced in late winter by excluding the dormant crown of the plant from light by placing a pot or forcing jar over it. A few weeks later there should be a crop of delicious, blanched white stems to harvest. Plants are usually discarded after being forced. Avoid forcing a plant in its first year as it will not have enough energy to grow properly.

RIGHT: Sea kale is common plant in the wild, particularly in coastal areas, but is also used in herbaceous borders as it has attractive foliage and flowers.

Storage: Sea kale's moment of glory in the culinary spotlight is short lived. Blanched stems should be eaten immediately after picking as they turn to mush in the freezer and will only keep in the fridge for a few days.

Preparation: Simply trim the stalks and wash well. The young external leaves are edible too but, to ensure they lose their bitter taste when you come to cooking/eating them they should be blanched beforehand.

Sea kale is one of the simplest plants to grow, requiring very little work for the aspiring gardener. Its luxuriant, attractive, blue-green foliage makes it equally at home in the herbaceous border and the kitchen garden. Food aficionados enjoy it best when the early spring, blanched stems are

lightly steamed and simply served up as a side dish with butter. As the name suggests, it originates from coastal regions, which explains its ability to cope with poor, arid conditions and exposed sites, and its ability to combat whatever the harsh elements throw at it. Yet, whilst the plant might be tough, beyond its hardy exterior lies a delicious treat providing one of the most sought-after delicacies appreciated by chefs and cooks worldwide. It is a native of Britain, again explaining how it can thrive in damp, maritime conditions, and it would have been harvested by the local ancestors way before it became a popular ornamental and edible crop for the garden. Nowadays, growing it at home is the only way to enjoy it – it is now a protected species, after the Victorian's passion for it exhausted the natural supply of this once abundant wild plant.

LEFT: Despite its attractive, delicate-looking flowers, sea kale is a tough herbaceous perennial and is suited to growing in wild, exposed sites, coping with both damp and free draining soil.

Not surprisingly, sea kale goes particularly well with fish dishes, but the blanched leafstalks or stems also can be eaten raw in salads or boiled and steamed. They can either be lightly steamed, where they retain their firmness with a slightly nutty flavour, or be reduced down to a spinachy texture, which tastes delicious when seasoned and flavoured with garlic and a squeeze of lemon to accompany salmon or sea trout.

There is just one commonly grown variety called 'Lilywhite', otherwise they appear in garden centres and catalogues simply as 'sea kale'. Plants can be difficult to source, so they can either be grown from seed, or even better they can be grown from root cuttings, sometimes called thongs, if a friend or neighbour has it in their garden.

Sea kale requires an open sunny position and free-draining soil, replicating the stony or sandy conditions it is used to in the wild. If the soil is heavy dig in plenty of grit or sand beforehand. Keep the area around the plant regularly weeded as it grows to avoid it becoming shaded or having to compete for nutrients. In autumn, it should be treated like any other perennial and cut down close to ground level once the stems and flowers start to fade.

Forcing sea kale

Although the foliage and stems of the plant can be harvested at any time of the year, the texture can be tough with a bitter flavour. The best way to enjoy sea kale is to force the plant into early growth by placing an upside down bucket, dustbin or a terracotta rhubarb forcing jar over the dormant crown in late winter. If using an old plant pot, then the drainage hole should be covered over to completely exclude the light. Place a brick on top of it to prevent it being blown away. A few weeks later there should be delicious pale white stems ready to harvest.

Use a knife to cut the stems just above the crown of the plant. Plants are often discarded after harvesting as they have expended their energy. However, do not to forget to take some root cuttings before you force as they will provide you with plants for the following year.

Cucumber and gherkin

Cucumis sativus

Common name: Cucumber and gherkin

Type: Climbing or trailing annual

Climate: Tender, cool or warm-temperate glasshouse

Size: 40cm, spreading 2m or more

Origin: South Asia

History: The cucumber was first cultivated in India more than 3,000 years ago and was brought to England by the Romans, although it did not become established until the 16th century. According to Pliny, the Emperor Tiberius had the cucumber on his table daily during summer and winter. The Romans reportedly used artificial methods (similar to the glasshouse system) of growing to have it available for his table every day of the year.

Growing gherkins

Gherkins are basically immature ridge cucumbers and are grown so that they can be pickled in jars of malt vinegar. Any ridge cucumber can be picked early although there are specific gherkin varieties that will perform better such as 'Diamant' and 'Venlo'. Harvest them when they reach about 6cm in length.

Cultivation: Indoor cucumbers should be sown between late winter and early spring, and outdoor cucumbers from mid to late spring. They need regular watering as they have such a high water content. Keep harvesting the fruits to encourage them to crop more.

Storage: Cucumbers do not store for long and will keep in the fridge for about two weeks. Harvest baby ridge cucumbers as gherkins and store them in jars of vinegar.

Preparation: When using in salads, keep the skin on and cut into thin slices, dice or chunks. Larger cucumbers or ridge cucumbers should be peeled thinly before eating.

BELOW: Ridge cucumbers are fairly hardy and can be grown outside in sheltered locations, but they will need a support system of wires and canes for the plants to climb up.

Cucumbers have a history dating back over 5,000 years and were originally enjoyed in India before spreading towards Europe and the remainder of Asia. No green salad is complete without the cooling and refreshing texture of a cucumber. Alternatively, mix them with radish, feta and nuts for a delicious crunchy salad. Lightly perfumed and very juicy, you almost drink the vegetable rather than eat it. Containing next to no calories, cucumbers are amazingly easy to grow, and yet strangely they are not that commonly grown in gardens. This is probably because they have a reputation for being tricky to grow, but the are actually surprisingly easy.

Not all cucumbers are long. There are lots of different varieties to grace people's gardens both outdoors and in the glasshouse. There are a range of different colours, including yellow and white, and round and oval shapes.

There are two types of cucumbers. Firstly, there are the climbing types that include the ones commonly seen in the supermarkets. They need to be grown under glass and varieties worth trying include 'Carmen' and 'Mini Munch'. The other type of cucumbers are known as ridge types, and these are hardier so can be grown directly outdoors without the need for protection. Varieties include 'Marketmore', which produces high yields and has a good trailing habit, and 'Tokyo Slicer', which produces long, smooth fruits. For something a bit different try the quirky heritage variety 'Crystal Apple'.

TASTING NOTES

Tzaziki

This traditional Greek and Turkish cucumber recipe uses just a few simple ingredients and is ideal as a dip or filling to go with pitta bread, or as a side dish for meat and cheese.

Preparation time: 10 minutes
Serves: 10 people (as a dip)

- 1/2 medium cucumber
- 2 tsp olive oil
- 150ml (5fl oz) natural yoghurt
- 1 tbsp mint, chopped
- 1 clove garlic, crushed
- Salt and pepper, to taste

Dice the cucumber, discarding any seeds, and place in a bowl.

Add the yoghurt and olive oil.

Next stir in the mint and garlic. Add salt and pepper to taste.

Cover and chill before serving.

RIGHT: Ridge cucumbers have slightly knobbly skins, which can be a bit tough and best if sliced off prior to eating.

LEFT: The smoother types of cucumber should be grown under glass to enable them to ripen. These are types commonly seen on supermarket shelve, but home-grown taste far better.

It is best to choose F1 cucumbers when growing indoor varieties as they should not produce male flowers. Male flowers will pollinate the female flowers and this results in bitter-tasting fruits. If male flowers do appear they should be removed immediately. To distinguish between the two flowers, look just below the flowerhead. The female has a swelling (which eventually goes on to form the fruit) whereas the male has nothing.

Growing indoor types

Indoor types should be sown in late winter in a heated glasshouse or in mid-spring in an unheated one. The large seeds should be sown on their edge, 1cm deep to prevent them rotting, then transferred into pots filled with potting compost when they reach about 25cm high. They require a bamboo cane or wire to train them upwards. Once the plant has reached the top, the growing tip should be pinched out. The tips of side shoots should also be pinched out two leaves past female flowers. Trim back any other flowerless shoots to about 45cm to prevent the plants expending valuable energy. Keep the plants regularly watered each day and feed with a liquid feed such as tomato fertilizer every 10 days from when they start to set fruit. The plants prefer humid conditions so damp the floor down once a day. It may be necessary to apply shade paint or hang a shade net up in the glasshouse to prevent the plants getting scorched from direct sunlight.

Growing outdoor ridge types

For outdoor types the seeds should be sown under glass in mid-spring before being hardened off in cold frames for a few days and then planted outdoors 75cm apart after the risk of frost is over. Alternatively, the seeds can be planted directly outdoors in early summer planting a seed 2.5cm deep every 75cm. A cloche can be placed over it to give it some initial protection during germination.

Cucumbers require a fertile, rich soil so add plenty of organic material prior to planting or sowing. The growing tip should be pinched out when it has produced about seven leaves to encourage the plant to form a bushy growth habit that will form lots of fruits. They can be trained up nets or tepees. The plants will need feeding about every 10 days with a liquid tomato feed. Keep the plants well watered and harvest the fruits regularly with a sharp knife every few days to ensure they continue to crop. Do not remove the male flowers on the outdoor types.

TYPES OF KITCHEN GARDENS

There are a plethora of different types of kitchen gardens. Vegetables do not have to be grown in conventional, regimented straight rows following a strict programme of crop rotation. Kitchen gardens can be modern and chic or rustic and informal. There is no right or wrong. Choosing a style or type of garden is a matter of personal taste and reflection of personality.

WALLED GARDENS

Traditionally, large stately homes would have had large kitchen gardens that would have grown crops to provide food for the owners of the house, their guests and their staff. They were usually grown in walled gardens as this provided extra protection from the elements, and imparted a warmer aspect enabling the gardeners to provide early crops to the kitchen and to extend the season into winter. Most people cannot emulate these grand gardens in their own back gardens, but they are great places to visit and get inspiration for what to grow.

POTAGER

This is a relaxed, informal type of garden that mixes ornamental plants with edible crops. It is named after the French word *potage* meaning soup, implying that the garden is a concoction of many different plants. Potagers usually have an artistic or creative element to the garden, where conventional crop rotation and formal straight rows are forsaken for aesthetic purposes.

For people with a small garden, mixing ornamental plants and vegetables may be the only option anyway. However, vegetables are beautiful in their own right. Onions provide stunning flowerheads, and brightly coloured cabbages and lettuce can form beautiful tapestries of texture. Wigwams with French and runner beans scrambling up them provide height and a splash of colour.

LEFT: Engraved frontispiece from Thomas Mawe (1760s–1770s) and John Abercrombie (1726–1806) *Every Man His Own Gardener* dated 1800.

ABOVE: The Czech painter Antos Frolka (1877–1935) painted many folk scenes, this one showing a lady growing vegetables.

ALLOTMENTS

For many people without a garden, small parcels of land can be rented, usually from the local council, for an annual fee. There is a wonderful social aspect to having an allotment and it is a great way to share ideas with like-minded allotmenteers. For others it is an escape from the house, and an excuse to enjoy the fresh air and a spot of healthy exercise.

For those people who do not think they have time to maintain an entire allotment, then it is possible to take on a half size plot, or do a 'plot share' with friends. There has been a huge revival of interest in allotments, and there are currently large waiting lists for available plots in most areas of the country.

COMMUNITY KITCHEN GARDENS

For those people without the time to maintain an allotment, it is possible to join community kitchen gardens, which are increasing in popularity and cropping up all over the country. The work in the garden is shared between a group of people, and usually the amount of time that is spent in the garden is then rewarded by the amount of fruit and vegetables taken home. Often surplus food or vegetables are distributed to local charities. It is not just individuals who take part in community kitchen gardens. Often schools, pre-schools and charities take part in these projects too.

LEFT: Allotments are the perfect solution for those people wanting to grow their own vegetables, but do not have a garden. However, with the recent increase in popularity of home-grown food, there are often long waiting lists for their availability.

'For all things produced in a garden, whether of salads or fruits, a poor man will eat better that has one of his own, than a rich man that has none.'

John Claudius Loudoun, *An Encyclopaedia of Gardening*, (1822)

Pumpkin and winter squash

Cucurbita maxima and *C. moschata*

Common name: Pumpkin, winter squash

Type: Climbing or trailing annual

Climate: Tender, cool glasshouse

Size: 40cm, spreading 2m or more

Origin: South and Central America

History: Pumpkins are the largest of the winter squashes. The word pumpkin originated from the Greek word *pepōn*, which means 'large melon'. The word was gradually morphed by the French, English and then Americans into 'pumpkin'.

It is said that Columbus carried pumpkin seeds back with him to Europe from America. Without pumpkins many of the early settlers in America might have died from starvation. The following poem is a testament to the pilgrims' dependence upon pumpkins for food:

For pottage and puddings
and custards and pies,
Our pumpkins and parsnips
are common supplies.
We have pumpkins at morning
and pumpkins at noon.
If it were not for pumpkins
we should be undoon.

Pilgrim verse, (c. 1633)

Nutrition

Pumpkin and winter squashes contain no saturated fats or cholesterol but are very rich in dietary fibre and anti-oxidant vitamins such as vitamins A, C and E. They are also a source of potassium, which can help to stabilize blood pressure.

LEFT: Pumpkins are easy to grow but need very fertile soil to make them fruit and swell to a decent size. Their seeds are also edible and delicious when lightly roasted.

Cultivation: Pumpkins should be sown indoors into small plastic pots in spring and planted out after the risk of frosts is over. They require a rich, fertile soil with plenty of organic matter. Dig lots of compost into the soil prior to planting. Seedlings should be planted at 1.8m apart and require regular watering and feeding as they grow.

Storage: In autumn, the fruits should be left in the sun for a few days for their skin to cure as this will enable them to store for longer. If the weather is wet, then they should be brought in immediately to prevent them rotting. Store them in a cool, dark place such as a garage, cellar or shed. The orange-skinned types will usually last for a few weeks in a frost-free location, whereas some of the blue-grey and green types will last right the way through winter. Chunks of pumpkin can be frozen, but the best method is to make them into delicious soups and pies and then stick them in the freezer.

Preparation: Cut in half and scoop out the seeds. Then cut into sections and peel and chop the flesh into even-sized pieces. Simply cook in boiling salted water for about 15 minutes or until tender. Alternatively pumpkin can be steamed or roasted. If roasting then treat like potatoes and add to the hot fat around a roasting joint. Pumpkins can also be stuffed and baked (like marrow) or used in soups and stews.

RIGHT: Pumpkins have a sprawling habit, providing lots of luxuriant growth. Give them plenty of space in the garden, planting them at least 1.8m apart. The flowers appear just as soon as the plant has got established.

Whether you crave sweet or savoury, there is a pumpkin recipe to please most people's palate. This chunky autumnal vegetable can be transformed into delicious food to sustain you throughout the following winter months, ranging from a thick warming soup to a delicious sweet pumpkin pie in a shell of filo pastry. It can also provide a tasty snack made from the chunky, crunchy seeds slammed in the oven for 20 minutes with a dash of paprika, nutmeg or cinnamon and splash of olive oil. Even if there is not a recipe to tickle your taste buds it can still appeal to your creative sides as everybody loves to carve a scary pumpkin face. Most people love the green-fingered challenge of trying to grow a huge pumpkin to impress friends and relatives in autumn. Finally if that doesn't satisfy you, then it can appeal to your practical side as the insides of the fruit can be scooped out, leaving the outer

BELOW: Pumpkins and winter squashes come in a wide range of shapes and colours, making them a beautiful ornamental addition either in the garden or when bought inside and dried.

skin to act as a bowl for your favourite autumnal dish. In the garden itself, pumpkins and winter squashes look resplendent with their impressive vines trailing through the vegetable patch and their brightly coloured skins looking translucent in the autumnal hues. To both the gardener and cook there is not really any difference between the pumpkin and the winter squash, and both are grown in exactly the same way.

Pumpkins and winter squashes should be started off indoors from mid to late spring in small plastic pots. Sow one seed per pot, placing it on its side to prevent it rotting when it is watered. Keep them on a sunny window sill or in the glasshouse until the risk of spring frosts has passed. They should then be hardened off in a porch or glasshouse for about a week before planting them outside.

Tasting Notes

Pumpkin pie

Capture the essence of autumn with this classic pumpkin dessert.

Preparation time: 15 minutes
Cooking time: 45 minutes
Serves: 6–8 people

- 175g (6oz) shortcrust pastry
- 450g (1lb) pumpkin, cooked
- 2 eggs, separated
- 5 tbsp caster sugar
- 150ml (¼ pint) milk
- Pinch of salt
- 1/4 tsp ground ginger
- 1/4 tsp grated nutmeg

Pre-heat a conventional oven to 200°C (400°F / gas mark 6 / fan 180°C).

Roll out the pastry, line a 18cm (7in) flan tin and prick with a fork. Cook for 10 minutes.

Sieve the cooked pumpkin to make 300ml (½ pint) purée.

Add the egg yolks and 3 tablespoons of sugar, then beat in the milk, salt and spices.

Turn into the pastry case. Bake for 40 minutes.

Whisk the egg whites until stiff. Fold in the remaining sugar. Spread over the top of pie.

Turn off the oven, place inside and leave until the meringue is lightly brown.

Pumpkins and winter squashes require a sunny site in fertile soil. The soil has to be very rich to enable the plant to provide enough sustenance for its long trailing habit, luxuriant growth and abundant fruit so that it can grow through to the end of the growing season. Such is the plant's hunger for fertile conditions that many gardeners grow it directly on top of the compost heap to maximize the amount of organic matter and moisture content that the plant craves.

Dig in lots of organic matter such as well-rotted horse manure or garden compost a few months before planting. Seedlings need lots of space to accommodate their rampant growth habit, so plant them 1.8m apart. As the plants establish they will need watering regularly and the area around them needs to be kept weed free. Some gardeners lay plastic down prior to planting and plant the pumpkins through it to help suppress the weeds and retain the moisture. When the plants start to flower, the plants should be fed with a potassium-based liquid fertilizer every 10 days , such as tomato feed, to help the fruit to set and develop their flavour and colour.

The fruits are ready to harvest in autumn when the foliage starts to die back and the pumpkins have reached their optimum size. Leave them to dry in the sun for a few days after harvesting to cure and harden their skin before bringing them indoors to either cook or carve, or both.

Growing a giant pumpkin

Not everybody can grow pumpkins to match the world record-breaking weight that currently stands at a staggering 2,009lb (911kg)! However, there are tricks of the trade that can produce fruits large enough to possibly win one of the popular 'largest pumpkin' competitions that happen all around the world at harvest festivals and autumn shows.

BELOW: A watercolour depicting a range of pumpkins and squashes dating from c.1800 and attributed to the Chinese artist and collector Wang Lui Chi.

Choose a large variety of pumpkin such as 'Prizewinner' or 'Dill's Atlantic Giant'. Give the plants lots of space and add plenty of organic matter into the soil. As the plant starts to grow, allow the first couple of fruit to develop, but remove the remainder as they will deprive the plant of essential nutrients. Feed it regularly with a liquid feed high in potassium. Once you can tell which one of the two remaining fruit is going to be the largest, remove the smaller one and place the remaining potential prize winner on a patio slab or a bed of straw to prevent it from rotting on the soil. With a bit of luck, the remaining fruit will go on to be a record-busting giant pumpkin.

Spaghetti squash

This unusual squash is a real hit with children and a brilliant way of getting them to eat vegetables without them realizing. It is so named because when it is cooked the fibres break down to form distinctive long strands that look like orange spaghetti. To many people on health drives it is considered to be a low-calorie and healthy alternative to pasta with just 40 calories per portion as opposed to a gut-busting 200. Demand outstrips supply in shops for this rare gourmet novelty when it comes into season in autumn, meaning that the only way to guarantee a supply of this dream nutritional snack is to grow your own. It can be baked, boiled, sautéed and even microwaved but the simplest way to cook the fibrous strands is to throw them in a wok and stir-fry them with red onions, elephant garlic and a pinch of paprika. Another option is to steam and serve with a bolognaise sauce and crusty garlic bread.

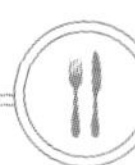

Tasting notes

Pumpkin and squash varieties to try

There is a huge range of different types of pumpkins and winter squashes to experiment with in the garden.

'Rouge Vif D'Etampes'	A flattened ball shape with a beautifully rich orange-red skin – the classic Cinderella pumpkin.
'Potimarron'	These small French heirloom types of winter squashes that are not much bigger than the size of a tennis ball.
'Crown Prince'	A smallish, quirky looking winter squash with steely blue skin, colourful rich orange flesh and a sweet and nutty flavour.
'Turk's Turban'	Uniquely shaped squashes with stripy green, orange and white skin.
'Butternut Harrier'	A classic with that distinctive large peanut shape ideal for cooking with rice to make a winter-warming risotto or simply roasted with seasonal herbs.

'Well, there doesn't seem to be anything else for an ex-president to do but go into the country and raise pumpkins.'

Chester A. Arthur, 21st President of United States, (1882)

Courgette, marrow and summer squash

Cucurbita pepo

Common name: Courgette, marrow, zucchini, summer squash

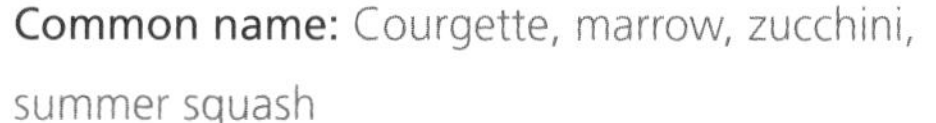

Type: Climbing or trailing annual

Climate: Tender, cool glasshouse

Size: 50cm, spreading 1.5m or more

Origin: Mexico

History: Courgettes are believed to have originated in Mexico about 7,000 years ago and archaeologists have traced their development in cultivation from between 7000 to 5500 BC. About 500 years ago courgettes were brought from Mexico to the Mediterranean by Christopher Columbus during one of his voyages. Before the 20th century, the courgette was not a popular vegetable in Europe or the United States, but now it is widely recognized in kitchens and home gardens.

BELOW: Courgettes are simply baby marrows, although varieties have been specifically bred that are more suited to being grown smaller. The flowers can also be fried in batter and eaten as fritters.

Cultivation: Start plants in plastic pots, indoors in late spring. Plant them out only after the risks of the spring frosts is over. Trailing types should be planted 1.2m apart and bush marrow types require about 80cm between each plant. Keep them well watered and harvest courgettes frequently to encourage them to continue cropping.

Storage: Courgettes do not last for long and should be cooked soon after harvesting, although they will keep in the fridge for a few days. Alternatively, they can be frozen but they lose their firm consistency. Marrows can be stored for a few weeks in a cool, dry and frost-free location such as the garage or shed.

Preparation: Slice off both ends, wash the skin and slice or dice as required. Small courgettes can be steamed whole, sliced and boiled in salted water for about 5 minutes, coated in batter and deep fried or sautéed in a little butter for about 5 minutes.

Courgettes and marrows must be one of the most versatile vegetables in the culinary world. The flesh can be baked, puréed, fried or grilled and then whipped up into any type of dish whether it be creamy, spicy, sweet, crunchy, smooth or whatever takes your fancy. Roasting is a great alternative as it intensifies the flavours when the moisture evaporates and the natural sugars caramelize and the texture turns sweet and crunchy. Despite originating in North America, these fruits are associated very much with Mediterranean cuisine. On warm, balmy summer days nothing beats al-fresco dining while grilling or frying a few strips of herby zucchini courgettes on the barbecue. The attractive, large orange-yellow flowers are also edible and can be fried up in batter to create a side dish of fritters.

In the garden, courgettes and marrows look beautiful. The most common colour is green, but there are attractive yellow courgettes such as 'Gold Star' and 'Sunstripe' or the creamy light green colour of 'Alfresco'. There are attractive looking marrows too, such as 'Tiger Cross' with its unusual green and white stripes on its skin, and they are not always elongated: there are attractive ball-shaped courgettes too, such as 'Summer Ball'. Some of the summer squashes also add interest to the vegetable garden, such as the colourful mix of 'Patty Pan' with its quirky-shaped, scalloped fruits. These can be harvested throughout the season at various sizes from baby squash, for eating raw, to their mature size of 20cm across.

Botanically there is no difference between the two vegetables. Marrows are simply oversized courgettes. If a courgette is left to mature on a vine it will get big, although specific varieties often have qualities better suited to whether they are grown as one or the other. Seeds should be sown indoors in

Nutrition

They are rich in vitamins A and C, courgettes, marrows and summer squashes contain no saturated fats or cholesterol. They are a very good source of potassium and also contain moderate levels of the B-complex group of vitamins like thiamin and riboflavin and minerals like iron, manganese, phosphorus and zinc.

ABOVE: This beautiful, colourful plate demonstrates the wide range of colours, shapes and sizes of summer squashes, marrows and courgettes.

spring in plastic pots in general-purpose compost. Sow one seed per pot. As with all members of the squash family, it is better to place the large seed on its edge as otherwise it can be prone to rotting in the compost. Keep the pots on a sunny windowsill or glasshouse until the risk of late spring frosts is over. They should then be hardened off in a cold frame or the porch for a few days before being planted outdoors.

Courgettes and marrows like to be bathed in sunlight from dawn until dusk, so plant them in a sheltered and warm, sunny spot in the garden. They require a fertile, well-drained but moisture-retentive soil so lots of well-rotted organic matter should be added to the beds a few months before planting. Plants with trailing habits will be happy sprawling on the ground, but they can also be trained up fences, hedges and wigwam teepees. There are also bush varieties that produce the fruit from the central stem, and these require far less space, making them ideal for the smaller garden.

Courgettes, marrows and summer squashes can also be sown directly outdoors after the risk of frosts is over. Sow two seeds every 80cm for bush types or 1.2m for trailing types. A bell-shaped cloche or a transparent plastic container can be placed over the seedling for the first couple of weeks to speed up germination and give it protection. Thin the seed out to one plant per sowing station after germination.

Once the plants are fruiting check them over every couple of days and harvest courgettes and baby squashes regularly to keep them productive and to prevent the fruits from getting too big. Both types of plant are thirsty and should be watered regularly and feed every 10–14 days with a liquid tomato fertilizer. Keep them weed free although often their rampant habit and large leaves suppress them anyway. Pick courgettes when they are about 10cm long. Cut through the stem with secateurs about 2cm away from the fruit. Courgettes should be cooked immediately after picking as they do not keep for long, whereas marrows will keep for a few weeks if picked at the end of the season and left in a frost-free place either in net bags or in trays.

BELOW: Summer squashes are very easy to grow but should not be planted outside until after the risk of spring frosts has passed. Plant in full sun in fertile soil and keep them well watered.

Tasting notes

Courgette muffins

A great recipe for an abundant crop, these muffins freeze well. Their flavour is slightly savoury but the icing adds a light, sweet twist.

Preparation time: 15 minutes
Cooking time: 25 minutes
Serves: makes 12 muffins

- 1 orange, halved
- 50g (2oz) grated courgette
- 1 grated apple
- 1 egg
- 75g (3oz) butter, melted
- 300g (11oz) self-raising flour
- 1/2 tsp baking powder
- 1/2 tsp cinnamon
- 100g (4oz) golden caster sugar
- Optional: 1 tub soft cheese mixed with 3 tbsp icing sugar, to make icing

Pre-heat a conventional oven to 190°C (375°F / gas mark 5 / fan 170°C).

Squeeze the orange and add to the grated courgettes and apple in a bowl.

Stir in the egg and butter.

Sieve in the flour, baking powder and cinnamon.

Add the sugar, and mix well until combined.

Spoon the mixture into muffin tins. Bake in the oven for 20–25 minutes.

Cardoon

Cynara cardunculus

Common name: Cardoon

Type: Perennial

Climate: Hardy, cold winter

Size: 2.5m

Origin: Mediterranean

History: The cardoon has been cultivated in the Mediterranean region for thousands of years. Roman gardeners are thought to have been responsible for taming this thistle into a garden vegetable, and it was a Roman custom to dip tender, young cardoon stems in a simple sauce of warm olive oil and butter and eat them raw. Cardoons were very popular in the Victorian era in Britain but are now grown mainly for their striking silvery foliage in the ornamental garden.

Cultivation: Plant in a sunny, sheltered site in well-drained soil. Harvest the stems in late autumn. Cut the plant down to ground level in early spring like any other herbaceous perennial. Remove the flowerheads that form during the year as this depletes the plants energy. They are easy to grow in the garden and although they can be grown from seed, it is quicker to buy plants from the garden centre. A free alternative is to find a friend with this plant in their garden, as it can be divided in late autumn by removing offsets with a sharp spade.

ABOVE: Not only can the foliage of cardoons be blanched and eaten, but their flowerheads are beautiful, making them suitable for the herbaceous border as well as the kitchen garden.

Storage: The stems do not last long and should be used within a few days of harvesting. They can be frozen but will lose flavour and texture.

Preparation: Rinse stalks well, trim the ends and remove any strings from larger stalks, also discard any discoloured outer stalks. Cardoons have a slightly bitter aftertaste so soak in salted water for an hour or so before cooking. Rinse stalks in running water before preparing to cook to remove the salt. Next, cut the stalks into 7–10cm (3–4in) lengths and cook in boiling salted water for about 30 minutes or steam until tender. Exact cooking times will depend on the size of stalk pieces and method of cooking. The stalks may also be sautéed in a little butter or braised. They make a tasty alternative side dish for any meal.

This tall graceful plant with its large, purple thistle head and spiny, silvery foliage is one of the aristocrats of the traditional herbaceous border, yet it also provides a bounty for the dining table. Very closely related to the globe artichoke, cardoons are grown in the kitchen garden for their long, silvery grey stems or stalks. This delicacy is a popular vegetable on the continent in particularly France, Italy and Spain, but it is considered in the UK as more of an edible curiosity rather than something delicious. The stems are usually boiled, steamed or braised and taste great when served with a rich and strongly flavoured cheese sauce or added to a creamy gratin. They provide a nutritional side as they are high in potassium, calcium and iron; they can also be baked or roasted. In Piedmont the stems are eaten raw with the dipping sauce *bagna càuda*.

Cardoons are monsters growing upwards and outwards to at least 2m, so give them plenty of space in the vegetable garden. Plant them in a sunny, sheltered site in well-drained soil.

The stems should be blanched to enjoy eating them. By excluding the sunlight from the stems it reduces the bitterness and makes them tender and succulent. To blanch them, in late summer the stems should be gathered up into the centre of the plant and tied together with garden twine. Wear gloves to do this, as the stems can be quite vicious and prickly. Do it on a dry day as excess moisture will quickly rot the stems and foliage. Wrap a hessian sack or cardboard around the stems and put a stout stake in place to prevent the upright column being blown over in the wind. Leave the cover in place for about three weeks and the reward will be delicious cardoon stems ready for the kitchen. Cut the remainder of the plant down and repeat the process the following year. Avoid blanching the stems on newly planted cardoons.

ABOVE: Cardoons have a long history of being cultivated for their ornamental qualities as they have beautiful architectural foliage and attractive blue thistle-like flowerheads. In the veg garden however, they should not be allowed to flower.

Nutrition

Cardoons are a good source of potassium, calcium, manganese, magnesium, copper and folates. They are free from cholesterol, fat and saturated fat. They also contain high levels of vitamin A, vitamin C, iron and phosphorus, and are a very good source of dietary fibre.

'A large garden vegetable in the luxury class, and not for small-space growing, the cardoon is as thistly in aspect as the globe artichoke.'

Charles Boff, *How to Grow and Produce Your Own Food,* (1946)

Globe artichoke

Cynara cardunculus Scolymus Group

Common name: Globe artichoke, artichoke, French artichoke

Type: Perennial

Climate: Hardy, average to cold winter

Size: 2.5m

Origin: North Africa

History: The globe artichoke was first cultivated over 3,000 years ago in the Middle East and was popular in the kitchens of ancient Rome. It was believed to be a potent aphrodisiac by the Romans. There is a Greek myth that the first artichoke was a woman of amazing beauty named Cynara, who lived on the island of Zinari. The god Zeus, who was there visiting his brother Poseidon, fell in love with her and decided to make her a goddess. Cynara missed her home and mother so much that she would sneak back to earth from Mount Olympus to visit her there. This infuriated Zeus, who in a fit of rage exacted his retribution by hurling her back to earth and transforming her into the first artichoke. It is from her name that we now get the botanical name for artichoke, *Cynara*.

Cultivation: Globe artichokes are best grown from plants or offsets as seed can give variable results. Plants should be spaced 75cm apart. They require a sunny, fertile soil. Add grit in the soil to make it less heavy. Mulch around the plant in spring to suppress weeds and keep the plants well watered. Harvest the flowerheads in summer when they reach the size of a tennis ball and before the flower opens.

Storage: Store globe artichokes in the fridge for a couple of days at the most because they do not last. Alternatively, they can be frozen but the stem and hearts should be removed. The hearts also can be stored in jars of oil or vinegar.

Preparation: Before cooking, cut the stalks off the artichokes and snip off a few of the rough outer leaves with scissors to remove any brown edges. Trim the tips of the remaining leaves.

LEFT: Globe artichokes are closely related to cardoons, but are grown for their edible flowerheads, which should be removed at this stage before they open.

If ever there was a delicacy that epitomizes a gourmet feast fit for a food connoisseur then this is it. Being regarded as a 'posh crop' by most people due to the small yield from each bulky plant, once the base of the edible flower bud or 'heart' has been tasted it is something that keeps chefs coming back for more, and it is not just foodies who love this plant – its beautiful, steely-grey ornamental foliage, statuesque purple flowerheads and wonderful ornamental and architectural structure mean it is adored by ornamental gardeners too. If left unharvested the wildlife also enjoy the heads as they attract a whole range of bees, butterflies and other beneficial pollinators to the garden. Varieties of note include 'Gros Vert de Lâon' for its superb flavour and 'Carciofo Violetto Precoce', although this variety is not as frost hardy.

The globe artichoke flowerhead is usually boiled or steamed and then two parts of the plant are eaten, the main part being the plump green or purple-tinged young flower buds collectively known as the heart. The other edible part is the base of the scale-like bracts. After boiling, these taste delicious when dipped in a hollandaise sauce or garlic butter, with the remainder of the bracts being discarded. Other popular methods of using artichokes include to chargrill them and add them to salads and risottos.

Globe artichokes require a fertile and well-drained soil in full sun. Because the plants are perennial, they will be in the same ground for a long time so it is important that the ground is prepared thoroughly prior to planting. Dig over the beds and

ABOVE: This decorative lithography is by Anton Seder (1850 to 1916), Art Noveau Print, called artichokes, clearly shows the decorative qualities of this attractive perennial.

BELOW: The edible part of the globe artichoke is the flower bud or heart in the centre of the flowerhead. The base of the immature scale bracts can also be pulled off and eaten.

Nutrition

Globe artichokes are an excellent source of dietary fibre, magnesium, manganese, niacin, riboflavin, thiamin, vitamin A and potassium. They are also a very good source of vitamin C and folic acid.

> 'Life is like eating artichokes;
> you have got to go through so much
> to get so little.'
>
> **Thomas Aloysius (Tad) Dorgan, cartoonist**

break up any hard, compacted soil that may be below the surface. Remove any perennial weeds in their entirety and then dig in plenty of well-rotted organic matter. Artichokes are often propagated from suckers or side shoots, sometimes called 'offsets', which are removed from the main plant in spring and potted up. Artichokes can be grown from seed in spring or autumn but this does produce variable results, so it is best to buy small plants or propagate your own from offsets. Seeds should be sown in a seed tray at a depth of 2cm. They should be potted on individually to 9cm pots a few weeks later and planted out in late spring. Plants should be spaced at least 75cm apart.

The globes are harvested with secateurs when they are about the size of a tennis ball, before they open and start to flower. There is often a second flush of flowerheads after harvesting and these too can be harvested and eaten in the same way, or left to develop and open for ornamental effect. In autumn the plant should be cut back to near ground level. Mulch around the plant each spring with a thick layer of mulch to suppress any weeds. Some of the less hardy varieties will need protection during the winter months.

RIGHT: Globe artichokes should be harvested when the head is about tennis ball size. They are ready for picking from mid to late summer.

Tasting notes

Baked globe artichokes

Once baked, this classic starter dish is delicious dipped in mayonnaise or added to salads.

Preparation time: 5 minutes
Cooking time: 50 minutes
Serves: 6 people (as a starter)

- 6 small globe artichokes
- 6 tbsp olive oil
- Salt and pepper, to taste

Pre-heat a conventional oven to 200°C (400°F / gas mark 6 / fan 180°C).

Trim the artichoke stalks.

Add the artichokes to a pan of boiled, salted water. Cover and simmer for about 30 minutes.

Drain and halve the artichokes lengthways and remove the heart with a spoon.

Place the artichokes cut side up on a baking tray and drizzle with olive oil. Season and bake for about 15–20 minutes.

Dahlia tuber (yam)

Dahlia

Common name: Dahlia, yam

Type: Tuberous perennial

Climate: Tender to half-hardy, cool glasshouse or mild winter

Size: 1.2m

Origin: Mexico

History: In 1525 the Spaniards reported finding dahlias growing in Mexico, but the earliest known description is by Francisco Hernández, physician to Philip II, who was ordered to visit Mexico in 1570 to study the 'natural products of that country'. They were used as a source of food by the indigenous peoples, and were both gathered in the wild and cultivated. It is believed that the Aztecs used them to treat epilepsy. Plants were taken back to Europe by Spanish adventurers more than 200 years ago. During the 1800s the popularity of dahlias surged; thousands of ornamental varieties emerged and were documented. After a brief time using dahlia tubers as a food crop to supplement potatoes, it was decided that they were better suited to decoration than food.

Cultivation: Plant tubers 10cm deep in the soil in the spring. If growing them to harvest tubers later in the year, most of the flowerheads should be removed to allow the energy to be channelled into the root system. Harvest the tubers in autumn when the first frosts have blackened the foliage.

Storage: Dahlias will store over winter if kept in a cool, dark, frost-free place. Cut back the stems and place the remainder of the plant in storage in boxes of sand or compost until ready to plant out again in spring. Check them over regularly for rot and remove those affected immediately.

Preparation: Wash and peel, the tubers, then dice. To cook, simply place in salted boiling water for about 20 minutes or until tender. They can also be roasted, baked or fried like ordinary potatoes.

BELOW: Dahlias originate from South America and were introduced to the UK originally as a possible blight-resistant alternative to the humble potato.

Putting the yum into yam, dahlias can be cooked just like a potato or a sweet potato to make sumptuous dishes. Considered by most gardeners to be a solely ornamental plant suitable for bedding displays or the herbaceous border, dahlia species were originally cultivated by the Aztecs for their tubers as an edible and nutritional food source. The dahlia species was introduced to the Western world by the 18th-century botanist Anders Dahl (hence the name dahlia), who considered the possibility that it would supersede the potato as part of the staple diet. Nowadays, dahlias are more of a curiosity than an edible treat and a fantastic talking point at the dinner table. Of course, dahlias that are bought from garden centres and seed catalogues have been bred for their flowering ability, and not for their taste. Therefore, not surprisingly, many varieties of tubers can often be small, watery and lacking in flavour. However, some of the older heritage varieties are closer to the original South American yam and are well worth a go.

Flavours vary and are often described as being nutty and similar to a water chestnut and the texture is slightly crunchier than a standard potato. They can be baked simply by being scrubbed and popped in the oven. If they taste bland, they can be jazzed up with creamy and garlic sauces or sliced and made into gratins. They can be made into crisps or chips or roasted.

RIGHT: The edible part of the dahlias are the tubers of the root system. These should be dug up in autumn, when the plumper ones can be removed for cooking.

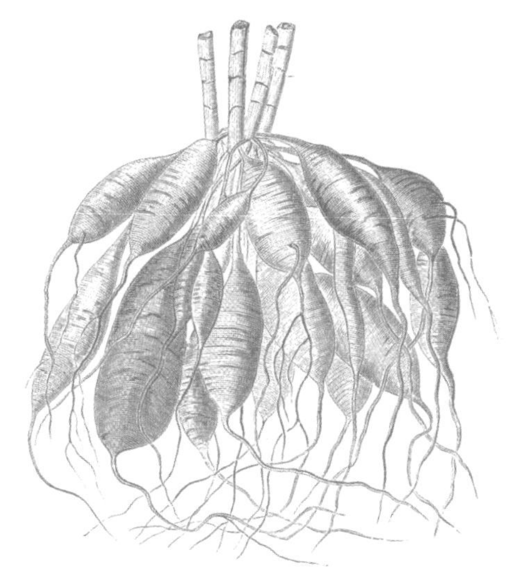

ABOVE: Dahlia tubers that are not to be eaten should be stored in a cool, dark and frost-free place and replanted outside in spring once the risk of frosts is over.

If you suffer from potato blight each year in the garden, then this could be the tuber for you. Avoid eating dahlia tubers that have been bought that year from the garden centre as they will probably have been chemically treated. Tubers should be planted in spring after the risk of autumn frosts is over. They require a fertile soil in full sun. Add lots of organic matter prior to planting and as the plants grow they may need support with stakes to prevent them blowing over. Tubers should be harvested in autumn when the foliage starts to die back. Dig the plants up carefully with a fork and remove about one third of the firmer, plumper tubers. Cut back the foliage and stems to about 10cm away from the root system. Place unused tubers upside down in sand and store them in a frost-free place over winter. Plant them out the following spring.

Carrot
Daucus carota

Common name: Carrot

Type: Annual

Climate: Half-hardy to hardy, mild to average winter

Size: 15–20cm

Origin: Middle Asia (Afghanistan)

History: It is believed that the carrot originated as a cultivated plant some 5,000 years ago in Middle Asia around Afghanistan, and slowly spreading into the Mediterranean area. The first carrots were mainly purple, with some white or black – not orange. Their roots were thin and turnip shaped. It is believed the ancient Greeks and Romans cultivated carrots. For example, carrots were mentioned in the writing of Pliny the Elder and prized by the Emperor Tiberius. Temple drawings from Egypt in 2000 BC show a purple plant, which some Egyptologists believe to be a purple carrot.

Cultivation: Sow seeds thinly in shallow drills between March and July. The early and late seasons can be extended by growing the plants under cloches. Thin them out to 15cm apart and harvest about 9 to 12 weeks after sowing.

Storage: Carrots can be stored for a few months after picking if placed in boxes of sand or compost. Remove all foliage first. Alternatively, they can be frozen.

Preparation: Small new carrots should have their stalks removed and simply be scrubbed under cold water. Larger, older carrots should be peeled and have the ends cut off, then simply cut them lengthways into batons or slices. Carrots can be eaten raw or cooked in salted boiling water for about 20 minutes or until just tender, or steamed for about 30 minutes.

BELOW: Carrots prefer a light sandy soil in full sun. Soils that are stony or heavy will cause the roots to fork and twist.

Nutrition

Carrots are rich in anti-oxidants, vitamins A and C and any B-complex vitamins such as folic acid, vitamin B6 and thiamine. They are also a rich source of carotenes. Its anti-oxidant property helps the body protect from diseases and cancers by scavenging harmful free radicals.

Carrots lend themselves to both savoury and sweet dishes ranging from coleslaw to carrot cake. Boiled, steamed or caramelized and roasted, there are so many dishes this popular orange root vegetable makes an appearance in. Yet, strangely, the orange carrot is a relatively new kid on the block and prior to the 16th century carrots were white, purple and yellow and not an orange one was in sight. It is supposed to have been the Dutch growers that developed the orange colour, probably as a patriotic tribute to the House of Orange. It could now be said that carrots have returned to their roots, as there is a revival in some of the older varieties of coloured carrots. 'Purple Haze' is a popular purple one worth trying while 'Rainbow' provides a selection of a few of the different colours all in one seed packet.

Carrots require a fertile, deep but light loamy soil that is free of stones as these can hinder the development of the tap root and cause forking. Carrots can also struggle to penetrate heavy clay soils, so they are better grown in raised containers and deep window boxes if space is a problem.

ABOVE: Orange is a relatively new colour for these sweet root vegetables. Originally they were purple, white and yellow and not until the 16th and 17th centuries did orange ones appear.

LEFT: Carrots are biennial and if left in the ground for the second year, will produce an attractive flowerhead that can be used for garnishing dishes.

Alternatively, short 'Chantenay' types of carrot can be grown such as 'Carson' or the round, globe carrots called 'Parmex'.

There is a particular trend for baby carrots at the moment, and these are simply ones that have been harvested from the ground before they have had a chance to mature and are easy to grow at home.

Carrots can be sown as early as February under cloches if early varieties such as 'Early Nantes 2' are

Tasting notes

Carrots Vichy

This is a classic French recipe using tender juicy carrots coated in a rich buttery glaze that makes a perfect side dish to many main meals. Substitute orange carrots for other colours such as purple or yellow to really make this dish stand out. This dish was named after the French town of the same name.

Preparation time: 5 minutes
Cooking time: 20 minutes
Serves: 4 people

- 600g (20oz) carrots, sliced julienned
- Knob of butter
- 1 tsp molasses sugar
- Salt and pepper, to taste
- 350ml (12fl oz) chicken stock
- Handful of mixed herbs, chopped

Put the carrots into a pan with the butter, sugar, salt and pepper.

Add the chicken stock, bring to the boil and cover and cook for about 10 minutes.

Remove the lid and furiously boil until the liquid evaporates and has produced a glaze – after about 10 minutes.

Tip into a warm serving dish and sprinkle with mixed herbs, such as parsley, coriander or chives.

chosen. Otherwise they should be sown between March and July. Sow seed thinly in shallow drills about 1cm deep in rows 15cm apart. Seedlings should eventually be thinned out to about 15cm apart –do not forget to eat the thinnings. For growing baby carrots a spacing of about 5cm between each plant is sufficient. Avoid using old seed as it tends to go stale quickly, so buy new to ensure that it is fresh and viable. Make sowings regularly every few weeks so that there are carrots to harvest regularly throughout the season.

'Sowe carrets in your garden, and humbly praise God for them, as for a singular and great blessing.'

Richard Gardener, *Profitable Instructions for the Manuring, Sowing and Planting of Kitchen Gardens*, (1599)

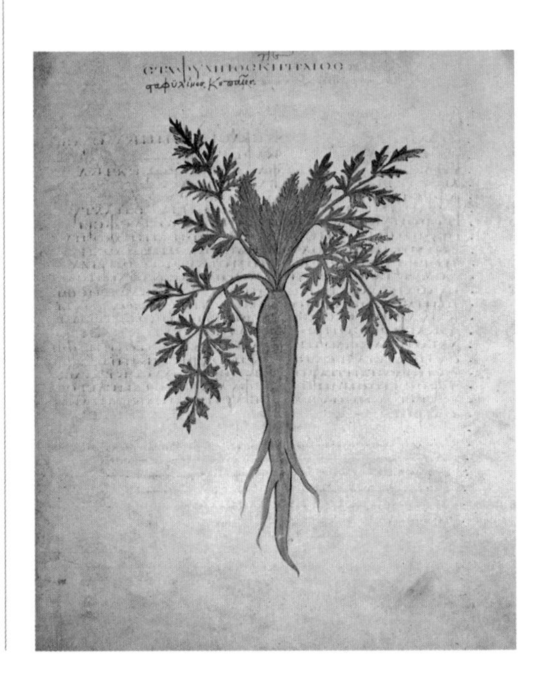

RIGHT: Carrots have been cultivated for centuries. This picture in the *Vienna Dioscorides*, an early 6th-century illuminated manuscript, shows a yellow variety of carrot.

ABOVE: Carrots have attractive feathery or fern-like, serrated foliage and pale white flat flowerheads. Wild carrots are commonly seen in the countryside and are said to have medicinal uses too.

If the root tops start to poke above the ground they should be covered back with soil to prevent them turning green. Due to their fine foliage they are fairly drought resistant and only in extremely hot and dry conditions do they need watering. Early varieties of carrot should be ready for harvesting about 9 weeks after sowing, whereas main crops such as 'Autumn King 2' will take about 12 weeks from sowing to maturity.

Main crop varieties will store for much longer than earlier varieties, but most can be stored in boxes in sand for a few months after harvesting.

Friends and enemies

Not only do carrots and onions go well in the kitchen in dishes such as coleslaw and relishes, but they also make great companions in the garden. Carrots can suffer from a pest called carrot fly, which lays eggs at the base of the plant; later, maggots hatch from the eggs and infest and ruin crops. It is thought that by planting onions next to carrots the flies are prevented from landing as they get confused by the aroma. As if that is not good enough, carrots also help to deter the onion fly pest.

Another solution to avoiding carrot fly is to grow resistant varieties such as 'Flyaway' and 'Resistafly'. Erecting a 60cm-tall mesh barrier is perhaps the most effective method as it excludes the flies, which only fly low to the ground.

WEEDING AND MAINTENANCE

Regular weekly weeding is vital if you are to get optimum growth from your vegetables, otherwise they will compete for water, nutrients, and in the case of the taller weeds – light. Left untended, annual weeds will start to self-sow and perennial weeds will rapidly spread, making the job a lot harder to get on top of. If beds are to remain empty for a while then it is worth placing a black landscape fabric over the surface to prevent the germination of weeds. Where weeds are a real problem on the plot, it is worth permanently retaining the fabric, and planting vegetables through it. Regularly mulching the beds should also suppress weeds and encourage the vegetables to grow and out-compete their competitors.

'I scarcely dare trust myself to speak of the weeds. They grow as if the devil was in them.'

Charles Dudley Warner, (1876)

ANNUAL WEEDS

Annual weeds usually have a small root system and can simply be hoed off. Work along the rows, moving backwards to avoid treading on areas already worked on. Push the hoe just below the surface of the soil, slicing through the roots and leaving them on the surface. The weeds can be left to desiccate in the sun during summer, but in less inclement weather they should be gathered up with a rake and added to the compost heap.

Annual weeds should be hoed off or pulled up before they set seed, otherwise they will rapidly spread throughout the vegetable plot.

Annual weeds include:

- Hairy bittercress
- Fat hen
- Shepherd's purse
- Goosegrass
- Annual meadow grass
- Groundsel

Annual weeds can also be burnt off using a flame gun, taking care not to damage the vegetables.

LEFT: Shepherd's purse is a commonly found annual weed which can be hoed off and added to the compost heap. Try to catch it before is releases its seeds from its purse-like pods.

PERENNIAL WEEDS

These weeds are the bane of the vegetable gardener's world. Once they have got their roots into the plot, they can be tricky to eradicate. Their root system can be invasive and quickly spread throughout the vegetable beds. Hoeing is ineffective as the deep, fleshy roots of perennials weeds will remain in the ground and quickly re-sprout new shoots.

For non-organic growers a systemic weedkiller can be used, which will kill the roots. Alternatively, the plants can be dug out using a fork. A spade can be used but care needs to be taken not to slice through the roots as this will help the plants spread yet further.

Fresh perennial weeds must not be added to the compost heap as they will rapidly spread. Instead, leave the weeds out in the sun to dry for a few weeks, which should kill the living material, which can then be added to the compost heap. If there is no sun forecast, some plants can be seeped in water or placed in black bags and left for a few months before being added to the heap.

Perennial weeds include:

- Bindweed
- Ground elder
- Perennial nettle
- Creeping buttercup
- Dock
- Spear thistle
- Speedwell
- Dandelions

Useful weeding tools

There are a variety of different tools that can be used for removing weeds from the ground.

Two-pronged weeding tools are useful for prising out perennials with long taproots. Ensure that each tine is on either side of the plant before lifting it out.

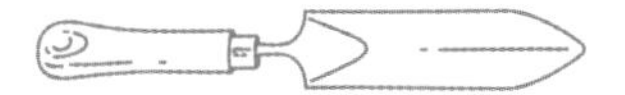

Narrow-bladed trowels can be used to lever out perennials or annuals in the vegetable bed without disturbing nearby roots

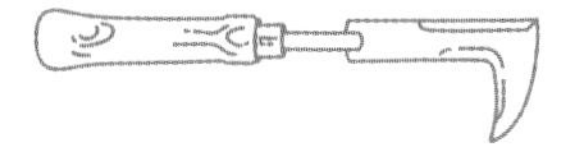

A patio-weeder tool is suitable for scraping between narrow gaps such as paving slabs or brick paths on the allotment, to remove weeds or moss.

BELOW: Bindweed is one of the worst types of perennial weeds to find in the herbaceous border as it is so invasive and gets tangled up among the plants.

Salad rocket

Eruca vesicaria subsp. *sativa*

ABOVE: Rocket is a very fast-growing, leafy vegetable and can be grown on its own, but is also commonly found in seed mixes for cut-and-come-again cultivation.

Common name: Salad rocket, rocket, rucola, Italian cress

Type: Annual

Climate: Half-hardy, mild winter

Size: 15cm

Origin: Mediterranean

History: In Roman times rocket was grown for both its leaves and its seed, and it was considered to be an aphrodisiac with added medicinal properties. It is one of the oldest vegetables cultivated and has been used in England in salads since Elizabethan times. Rocket has become popular again recently and features heavily in the Mediterranean diet.

Cultivation: Seeds should be sown in shallow drills from late summer through to mid-autumn. Plants need covering with cloches as winter approaches so that leaves can be regularly harvested throughout winter.

Storage: Leaves do not last for long once picked, so harvest regularly from the garden as and when needed.

Preparation: To prepare, simply pick the rocket leaves and wash thoroughly under cold running water to remove any dirt or pests.

Nutrition

Rocket is an excellent source of vitamins A, C and K and rich in the B-complex vitamins such as thiamin, riboflavin, niacin and vitamin B6. These vitamins help to promote a healthy immune system and good bone health.

There is a plethora of trendy salad leaves that are very much in vogue not just in restaurants but also in supermarkets where their shelves are packed full of vacuum-packed assortments. Leaves range from the bitter chicories and endives to the peppery Japanese brassicas such as mizuna and Chinese cabbage. Often these bags are very expensive and yet the leaves are so easy to grow at home. These plants hardly take up any space at all, and can even be sown in a window box just outside the kitchen so that handfuls can be cut and harvested and thrown into a dish without even having even to step out into the garden. One of the most popular salad leaves is rocket because it is so easy to grow, and it has a unique spicy and peppery flavour. The older leaves are slightly hotter and are usually steamed or added to stir-fries, make a superb spinach substitute. The younger leaves are usually

LEFT: Rocket can be grown for most of the year but can be prone to bolting in summer, so is often sown in autumn under cloches and treated as a winter and spring crop for salads.

Tasting notes

Simple rocket and Parmesan salad

This is a versatile dish that will go with many different flavours. You can even add chopped, crispy bacon for some extra crunch.

Preparation time: 5 minutes
Serves: 2 people (as a side dish)

- 1 large bunch rocket
- 45g (1½oz) Parmesan cheese
- 100ml (3fl oz) olive oil
- 4 tbsp lemon juice
- 2 tsp sea salt
- Black pepper, to taste

Wash the rocket and lightly pat dry.

Shave the Parmesan using a potato peeler.

Whisk together the olive oil, lemon juice and salt in a large non-metallic bowl.

Add the rocket leaves and toss together.

Simply arrange the dressed leaves on a serving plate. Scatter the Parmesan shavings over the rocket, add pepper and serve.

eaten raw in salads or even sandwich fillings. The flowers are also edible. The leaves are often grown as cut-and-come-again leaves, which simply means that the leaves are cut with scissors near the base of the plant and it re-sprouts, providing numerous servings throughout the growing season.

Sow seeds regularly between April and September in fertile, well-drained soil or in a window box. Plants should be thinned out to a spacing of between 15 and 20cm. As the name suggests, rocket is a fast grower and it is possible to start harvesting leaves just three weeks after sowing. Varieties worth trying include 'Apollo' and 'Runway'.

Growing for winter

Rocket is a great autumn and winter salad crop when lettuce crops are scarce. It loves it slightly cool to bring out the best of its flavours and will very often bolt and run to seed if the weather is too dry or hot. It should be sown thinly in late summer or early autumn in shallow drills 1cm deep in rows 30cm apart. After germination the rocket can be thinned to 15cm apart – remember to add the thinnings to the next salad. As cooler weather approaches, the plants should be covered with a cloche. Harvest throughout autumn and winter by either snipping off individual leaves or by harvesting the individual plants.

Other types of cut-and-come-again salad leaves

Rocket can be grown on its own or as part of a seed mix. Browsing through seed catalogues enables you to pick out mixes that appeal to your taste buds. Other popular types of cut-and-come-again include land cress (*Barbarea verna*), white mustard (*Brassica hirta*), garden cress (*Lepidium sativum*), winter purslane (*Claytonia perfoliata*) and lamb's lettuce (*Valerianella locusta*).

ABOVE: Garden cress (*Lepidium sativum*)

RIGHT: Rocket was grown by the Romans for both its aromatic leaves and its seeds. It was also considered to have medicinal qualities and was used as an aphrodisiac.

Florence fennel

Foeniculum vulgare var. *azoricum*

Common name: Florence fennel, finocchio or bulbing fennel

Type: Perennial

Climate: Tender, frost-free winter

Size: 60–80cm

Origin: Mediterranean

History: The word 'fennel' comes from the Middle English *fenel* or *fenyl*, meaning 'hay'. In Greek mythology, Prometheus used the stalk of a fennel plant to steal fire from the gods. Also, it was from the giant fennel that the Bacchanalian wands of the Greek god Dionysus and his followers were said to have come.

Cultivation: Sow seeds from late spring to early summer and thin them to 25cm apart in rows 35cm apart. Do not sow too early as they have a tendency to bolt.

Storage: Fennel bulbs will not last for long after being harvested, and can only remain in the fridge for a few days.

Preparation: Florence fennel's bulbous root can be blanched or eaten raw. Fennel is also good braised. To prepare, simply trim both the root and the stalk ends. Chop or grate if it is to be eaten raw. To cook, quarter the bulb and cook in boiling salted water for about 30 minutes. Drain and slice. The slices can then be sautéed in melted butter.

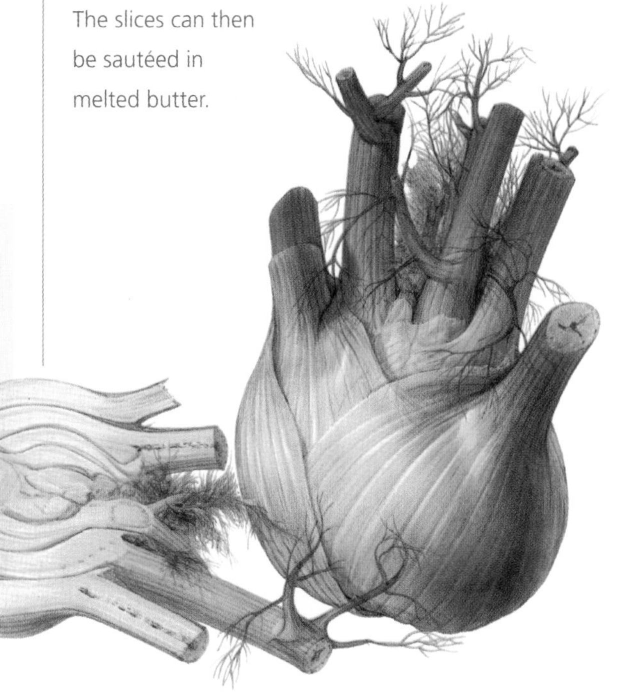

RIGHT: Contrary to popular belief, Florence fennel is not a bulb, but is a swollen stem. The attractive flowerheads produce seeds that can also as flavouring in cooking.

Nutrition

Fennel seeds and bulbs are a good source of dietary fibre and aid digestion. The seeds and bulbs are also a good source of vitamin C and other nutritious vitamins and minerals. Fennel also contains high levels of the B vitamin folate, which is especially helpful for maintaining the health of our blood vessels.

The name gives away the origins of this peculiar looking vegetable with a fat, swollen white stem at the base of the plant. Florence fennel, or finocchio as it is more commonly known in Italy, is very popular in their cuisine. It has a strong aniseed flavour that some may find overpowering when sliced or grated and served raw in salads, but when cooked the flavour becomes much milder. The vegetable goes particularly well with strong fish dishes such as salmon and sardines, and also combines well with celery and clean-tasting flavourings such as mint or lemon juice. The bulb is usually steamed or boiled but it can also be roasted whole. The strongly flavoured aniseed foliage can also be used for flavouring stews and fish soup or simply used as an attractive garnish.

Fennel requires a fertile, light soil in plenty of sunshine. On heavy clay soils, it may be better to grow in raised beds. Alternatively plenty of grit should be added to help with drainage. Seeds can be sown indoors in pots or modules but they dislike root disturbance so directly sowing outdoors is a better option. There is no rush to sow the seeds, because if sown too early they have a tendency to bolt, although there are bolt-resistant varieties worth trying such as 'Zefa Fino', 'Cantino' and 'Amigo'. Aim to sow thinly between mid-spring and early summer in 1cm drills in rows 35cm apart. Thin the plants to 25cm apart after germination.

BELOW: Florence fennel has a slight aniseed flavour, similar to star anise, and is often used to flavour fish dishes or poultry. The foliage can also be used for flavouring.

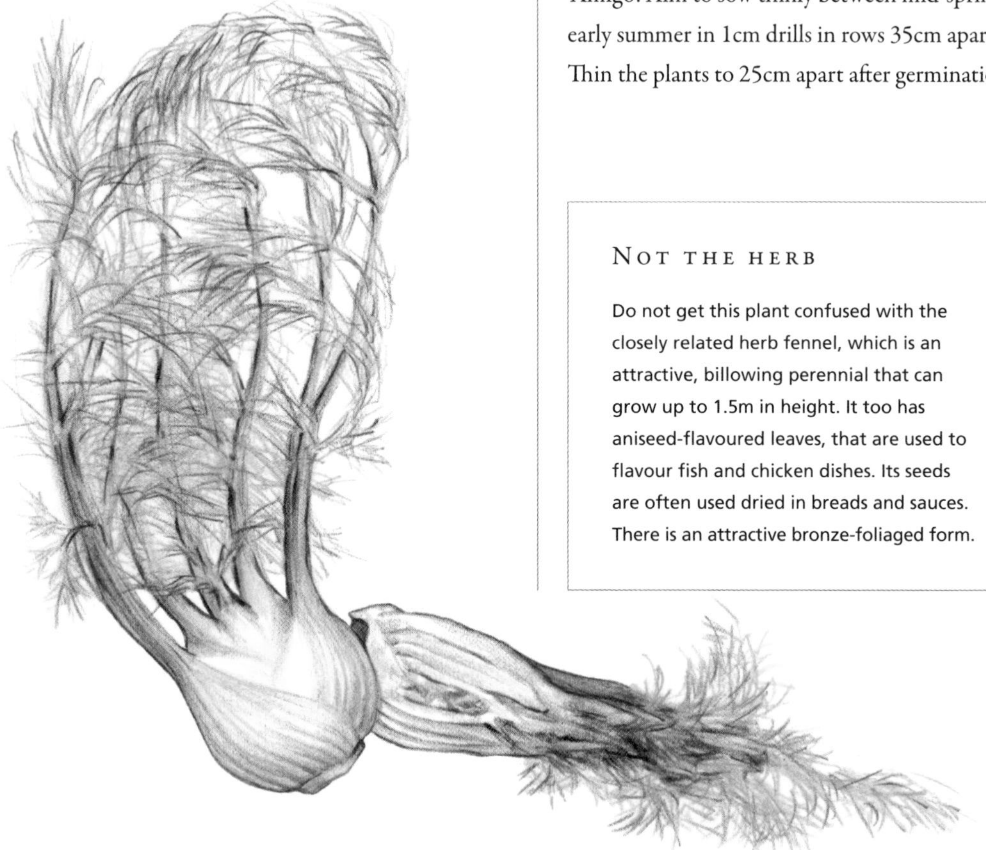

Not the herb

Do not get this plant confused with the closely related herb fennel, which is an attractive, billowing perennial that can grow up to 1.5m in height. It too has aniseed-flavoured leaves, that are used to flavour fish and chicken dishes. Its seeds are often used dried in breads and sauces. There is an attractive bronze-foliaged form.

ABOVE: The key ingredient to getting the stems to swell is plenty of water. If it is dry in springtime, the plants will need watering practically every day.

The two essential ingredients to getting a decent swollen stem from Florence fennel are water and sunshine. It needs lots of moisture to enable the stem to swell properly, so in dry spells it is essential that the plant is kept well watered. As the plant expands the lower section of the stem can be earthed up with soil in order to blanch it and sweeten up the flavour.

Bulbs are usually ready for harvesting about 14 to 16 weeks after sowing. The bulb should be between the size of a golf ball and tennis ball. Cut about 2cm above the ground as this will encourage a further flurry of feathery shoots to appear later, which can be harvested and used for aniseed flavouring and garnish.

Tasting notes

Simple roasted fennel

Roasting fennel stems brings out their inherent sweetness. Delicious if served with roasted lamb, chicken or fish dishes.

Preparation time: 5 minutes
Cooking time: 40 minutes
Serves: 4 people (as a side dish)

- 2 fennel bulbs, stalks cut off, bulbs halved lengthwise and cut lengthwise in thick pieces
- 2 tbsp olive oil
- 1 tbsp balsamic vinegar

Pre-heat a conventional oven to 200°C (400°F / gas mark 6 / fan 180°C).

Brush olive oil over the fennel pieces and sprinkle on some balsamic vinegar.

Grease a baking dish and lay out the pieces of fennel; roast for 30–40 minutes.

Jerusalem artichoke

Helianthus tuberosus

Common names: Jerusalem artichoke, sunroot, earth apple, sunchoke

Type: Tuberous perennial

Climate: Fully hardy

Size: Up to 2.5m

Origin: North America

History First cultivated by Native Americans, the French explorer Samuel de Champlain discovered plants cultivated at Cape Cod in 1605 and brought some back to France.

Mysterious name

Bizarrely, the Jerusalem artichoke has nothing to do with globe artichokes. The name is attributed to the French explorer Samuel de Champlain, who sent plant samples to France during his travels in the early 17th century, noting they had a similar flavour to globe artichokes. The vegetable also has nothing to do with Jerusalem. One theory behind its name is that it is a corrupted form of *girasole*, the Italian name for sunflower, to which it is closely related, as used by Italian settlers in North America.

By the mid 1600s, the Jerusalem artichoke had become a very common vegetable in Europe and reached its peak of popularity in the 19th century.

Cultivation: Plant tubers 5–10cm deep in fertile, well-drained soil. Harvest in autumn and winter as required.

Storage: Keep in the ground until ready to use. Avoid freezing as the texture deteriorates unless it's going to be puréed. After harvesting, store the tubers in a cool, dark, frost-free place, such as fridge or shed during winter until ready for cooking.

Preparation: Par-boil the tubers before attempting to peel them as this makes the skin come off more easily. The flesh rapidly discolours when exposed to air, so immediately place them in water with a dash of lemon juice after peeling or chopping.

Closely related to sunflowers, this impressive perennial vegetable provides a dazzling display of attractive, tall yellow flowers, yet its real treat lies buried below the surface. Its knobbly, reddish-brown tuber is a gourmet treat in the kitchen. It is expensive to buy in shops and yet is probably one of the easiest vegetables to grow in the kitchen garden. If you have a tendency towards

RIGHT: Jerusalem artichoke enriches both the garden and the veg plot thanks to its striking yellow flowers, which appear from midsummer.

laziness in the garden, then this has to be the perfect plant for you. It out-competes surrounding weeds and thrives in most soils, requiring hardly any maintenance at all. It is incredibly easy to grow – in fact, almost too easy. If the plant is not kept in check, it can become large and unruly and smother other nearby plants. However, it will reward you with a bumper crop each year for minimum effort.

Cook with them as you would the humble potato. They can be boiled, roasted, sautéed, baked and mashed. They have a distinctive nutty-yet-sweet flavour and unlike potatoes can also be sliced raw and added to salads or stir-fries. For total indulgence, try chopping them into chunks, dipping them in batter and frying them.

Tubers for planting can be bought from good-quality supermarkets and specific varieties can be purchased from seed companies.

LEFT: A delight for the gourmet gardener, Jerusalem artichoke is an expensive vegetable to buy in shops, but this plant is remarkably easy to grow and its reddish-brown tuber is simple to cook.

Carefully select where you are to grow them, ideally avoiding the south and west sides of your plot as their 3m height will cast shade onto the vegetable plants beyond. They should be planted out in early spring, burying each tuber 10cm deep and 50cm apart. Water the plants well after planting.

Tubers can just be left in the ground to regenerate each year. However, their quality will gradually deteriorate. Ideally the ground should be cleared every three or four years and the healthiest tubers selected and planted in a freshly dug patch. This is also a good way of keeping their spread in check.

Once the plants are about knee-height, pull the soil up around the base of each stem to about 15cm to prevent them swaying about in the wind. As the foliage starts to turn yellow in autumn, the stems can be cut back to just above ground level. Lay the stems over the soil to keep the frost off the ground, making them easier to harvest during the colder months. Tubers are ready for harvesting in autumn and can remain in the ground throughout winter. Dig them up using a fork, taking care not to spear the tubers.

Nutritional benefits

The tubers are very high in iron, vitamin C and both phosphorous and potassium. They also contain high levels of inulin, which is associated with intestinal health because of its probiotic properties or beneficial bacteria. However, do not overindulge with artichokes if you have a slightly sensitive stomach or a tendency for flatulence. Their effect on the body's gastric wind output is legendary!

Daylily
Hemerocallis

Common name: Daylily

Type: Rhizomatous perennial

Climate: Hardy, very cold winter

Size: 1m

Origin: Far East

History: Daylilies were valued throughout the Orient for medicine and food as well as for their beauty. They appear in a lot of Oriental art and legends. Daylilies arrived in Europe from China, Japan, Korea and Eastern Siberia during the 16th century, and by the 17th century had crossed the Atlantic to North America.

Cultivation: They need very little looking after. Keep them weed free during the summer and watered in dry periods. Cut the flowered stems back down at the end of the growing season. They prefer full sun but will tolerate some shade. Divide the plant every few years by slicing through the rhizome root system.

Storage: Flowers do not last for long, but can be dried, which will extend their culinary life for a few days. Roots will store for a few weeks if kept in a cool, dark place such as the garage.

Preparation: To prepare simply cut the 2.5cm (1in) long rhizomes from the roots and wash in cold water to remove the dirt. Boil in salted water for 15 minutes.

LEFT: Daylilies are a beautiful addition to the garden, with striking coloured flowers coming in an array of colours including red, orange and yellow. They are very easy to grow and low maintenance.

Daylilies are literally one of the perennial favourites in the garden for their gloriously coloured flowers coming in fiery, reds, oranges and yellows and their attractive, long strap-like leaves. However, unbeknown to many the entire plant is edible, stretching from the tallest leaf right down to its root system, making it not just a feast for the eyes but a feast for the stomach too. They have been grown for centuries in Asia as a food crop, particularly in China and Japan, and yet despite the worldwide popularity of daylilies as an ornamental plant, many people are missing out on a golden opportunity to sample an array of gourmet delights.

Daylilies' beauty is fleeting and in fact the botanical name *Hemerocallis* means 'beautiful for a day'. Individual blooms last for one day, opening in the morning and fading at night, hence their name daylilies. However, this unusual flowering phenomenon means that there are always plenty of tasty swollen, plump flower buds loaded up and ready to be harvested and used as a substitute any recipe calling for green beans. Do not harvest all of the flower buds though as you want some of them to produce flowers, which are the highlight of the plant's edible prowess. These brightly coloured blooms add a vibrancy and peppery zing to any salad. Use them for garnishing platters too. Alternatively they can be fried up in batter, a bit like a courgette flower, and eaten as a side dish.

Tasting notes

Edible parts of a daylily

Daylily offers a part of its plant for cooking almost all year round, including winter when part of the roots can be harvested.

Young shoots	Add emerging shoots to stir-fries and salads in spring.
Roots	Boil or bake them like a potato all year round, though they are best in autumn.
Leaves	Try in green salads and stir-fries in early to late summer.
Flower buds	Add to stir-fries in summer.
Flowers	Use in salads, stir-fried or cooked in batter in summer.

Fresh flowers can also be stuffed with dried fruits, nuts and cottage cheese with sweet herbs, tying the ends of the flowers together with chives stalks. The flowers have a slightly unusual chewy, thick texture that sets them apart from many other flowers. They can be dried and added to salads or stir-fries

The fresh emerging shoots have a very mild oniony flavour and can also be harvested, and can be chopped and added to stir-fries, or sautéed or steamed anytime in spring. Select shoots no longer than 12cm tall as anything bigger will be too coarse.

LEFT: The flower petals are delicious when added to salads, but the swollen buds are equally tasty and can be used as a substitute for green beans in most recipes.

Very tender shoots can be used raw in salads or even as fillings for sandwiches. Daylilies require some sun to produce their abundant flower display but they will tolerate some dappled shade during the day. They require very little maintenance, not usually requiring staking or deadheading. They are simply cut back after the foliage starts to die back, but in milder areas its attractive strap-like leaves will remain above the ground. Every three or four years the plant should be lifted and divided, by slicing the plant into pieces with a sharp spade, throwing away the congested centre and replanted the fresher sections.

The best way to get daylilies growing in the garden is to buy young plants from the garden centre and plant them in a sunny or partially shady location in spring. Alternatively, find a friend with a clump of it growing in their garden and slice through the root with a spade and take a section of it. It will not do their plant any harm at all. After planting simply water the plant in and give the area around the root system a thorough mulching with organic matter such as garden compost.

In addition to dividing the plant every three or four years as part of the plant's general maintenance, it is possible at the end of each season to dig the plant up gently and harvest some of the roots. Daylilies are robust enough to tolerate this and once a few roots have been removed for the kitchen the plants can simply be planted back their holes ready to flower again next year. There is no need to peel them; just give them a scrub and either bake or boil them like you would a potato or Jerusalem artichoke. Their flavour is reminiscent of turnips and nuts. They are recommended for detoxification diets.

Nutrition

Daylily flowers and tubers are high in protein and oils. The flower buds are good sources of beta carotene and vitamin C. They are very good in detoxifying the entire body system.

BELOW: The name *Hemerocallis* means 'beautiful for a day'. Each flower lives and dies within the day, but thankfully they keep on producing flowers for most of midsummer.

Plantain lily
Hosta

Common name: Hosta green shoots, plantain lilies, giboshi (in Japan), urui

Type: Perennial

Climate: Very hardy, very cold winter

Size: 65cm

Origin: Japan, China and Korea

History: Most of the species of hostas that provide the modern plants were introduced from Japan to Europe by Philipp Franz von Siebold in the mid 19th century.

Cultivation: Ideally hostas should be planted in shade in moist soil. However, they will tolerate sunshine. Keep them well watered during the year and watch out for slugs, which will quickly munch through the emerging shoots. Cut back the plant in autumn and mulch around it in spring to help retain the moisture.

Storage: Hosta shoots will not last for long and the leaves quickly turn limp once they are picked. Harvest them from the garden in spring and early summer as and when they are needed.

Preparation: Hosta green shoots are best cut when young and tender, preferably in the early morning, and eaten as soon as possible. Similar in taste and texture to asparagus they should be cut with a paring knife close to the soil. Then they need to be rinsed well under cold water. The stems can be eaten raw or cooked but they are more bitter when taken from mature plants. If cooked place in salted boiling water and boil for about 3–4 minutes. Serve hot smothered in melted butter and cracked black pepper.

Hostas are usually grown for their ornamental qualities with their showy, large foliage and long extended flower shoots. Perfect for woodland gardens and damp, shady corners, this popular herbaceous plant has yet another quality lurking amongst its attractive undergrowth. The baby shoots and leaves that emerge from the damp woodland soil are a popular springtime treat in Japan, where they are known as *urui*. The flavour has a taste reminiscent of a concoction of asparagus, lettuce and spinach and they can be eaten raw or cooked. In addition the flowers are edible and can be added to salads or used as garnish.

There are various ways of using hostas in the kitchen, where their mild but crunchy texture can be used to enhance and complement a range of stronger flavours. The shoots and leaves are usually boiled or blanched and then added to salads and stir-fries. They work well with the sweetness of red onions, garlic and fried tomatoes drizzled in honey or a Japanese sake wine dressing. They can also be fried up in batter as tempura and eaten dipped in a sweet plum sauce. Try them in wraps with a sweet and sour dressing and a goat's cheese filling with dried cranberries, or alternatively cook them up in omelettes with mushrooms, peppers and sweet herbs. They can also be steamed in rice and wrapped up in seaweed or nori for homemade sushi.

Urui is incredibly hard to get hold of in most shops and the only guaranteed way of a fresh harvest each year is to grow your own.

All hosta species appear to be edible, but *H. montana* and *H. sieboldii* are the most commonly used kitchen favourites as not only are they easy to grow but their leaves and shoots are reputed to have the best flavour with the least amount of bitterness.

Growing hostas

Hostas are associated with damp, woodland gardens but they will also tolerate sunnier locations and can be grown in an herbaceous border or even the vegetable patch, though it is important to keep the base of the plant moist. Prior to planting dig over the soil and add plenty of organic matter, which will help retain the moisture. If possible, add a barrow-load of leaf mould as this helps to replicate the conditions of a forest floor that hostas originally dwell in back in their native Asian woodlands. Plant hostas at about 60cm apart depending on variety. Select a handful of plants if you plan on harvesting the young shoots, as just one plant will not provide much *urui* each year. Water the plant in thoroughly after planting. During the season keep the areas weed free to avoid competition for moisture and keep an eye out for slugs and snails, which feast on the shoots (see p.218).

Every two or three years hostas benefit from being dug up and having their rootball divided, by ruthlessly slicing through it with a sharp spade. Discard the central congested section and replant the remaining fresh sections back in the soil. This not only reinvigorates the plants, but also

ABOVE: Hostas are the perfect ornamental foliage plant for a shady area of the garden, but they are also a gourmet treat and in Japan are a very popular delicacy. They flower in summer.

multiplies your stock, meaning in subsequent years there will be even more urui to harvest.

When the shoots emerge, resist the temptation to pick all of them as this will kill the plant. Remove no more than a third, although several pickings over time can be made from the plant as it comes into growth. It is generally considered better to harvest early in the morning as the moisture content is higher. Shoots should ideally be between 12 and 15cm long, but it is better to catch them while they are still rolled and have not unfurled, although larger leaves can be picked and cooked as a substitute for spinach. The summer flowers are pretty but very short lived.

Sweet potato
Ipomoea batatas

Common name: Sweet potato, Spanish potato

Type: Tuberous annual

Climate: Tender, frost-free winter

Size: 40cm

Origin: Central and South America

History: Despite its name the sweet potato is not related to the common potato. Its history dates back to 750 BC in Peruvian records and it was eaten in Europe well before the true potato. In fact, Columbus brought the sweet potato to England from the island of Saint Thomas in 1493. The Tudors considered them to be an aphrodisiac. The Spanish word for sweet potato is *batata* and in French it is *patate douce*. In the 1490s the English term 'potato' referred to the sweet potato rather the generic white potato of today.

Cultivation: Sweet potatoes like a fertile, well-dug soil in full sun. The ground should be cleared of weeds and dug over, incorporating lots of well-rotted manure. Place black landscape fabric on the ground to help retain warmth and moisture and suppress weeds and then cut holes in it, to plant the mature slips through at 30cm apart with 75cm between rows. They should be planted deeply, partly covering the stems with

ABOVE: Sweet potatoes are harvested from the ground like common potatoes, but are completely unrelated. In this image, it uses an orange-flowering plant as a support to help it grow.

Nutrition

Sweet potatoes are low in sodium and very low in saturated fats and cholesterol. They are also a good source of dietary fibre and contain good levels of vitamin A, C and B6, as well as potassium.

soil as this will encourage a larger crop. They do require a good growing season, so to help them on their way place a cloche over them, and water them regularly as they grow. If growing under a cloche, then this will need ventilating on warm days.

The tubers form underground and are harvested in a similar way to potatoes. After 12–16 weeks the foliage will start to die back, a sign that they are ready for harvesting. Lift them with a fork before the first frosts, and enjoy.

Storage: They will store for a few months if their skins are cured in the sun or glasshouse and then kept in a frost-free place, but can also be used immediately after harvesting.

Preparation: Scrub the sweet potatoes well and, if boiling, peel after they are cooked as the flesh is soft and rather floury. They can be boiled, baked, fried or roasted like the potato.

These gourmet potatoes must be one of the most versatile vegetables available, being a sweeter alternative to the humble spud. They can be used in so many dishes and are suitable for frying, boiling, roasting and mashing, and they make the most amazing chips and crisps. Try using mashed sweet potato as an alternative topping to cottage pie and you will never look back.

They can be grated and eaten raw, and even their leaves and stems can be picked and added to

Sweet potato family

Despite their culinary similarities to the potato, the sweet potato *Ipomoea batatas* is not actually related to the potato family, but is instead from the same group as the bindweed *Convolvulus* and the beautiful ornamental climber, morning glory. Thankfully sweet potatoes lack the vigour of their cousins, and although they have a sprawling habit they are much more manageable.

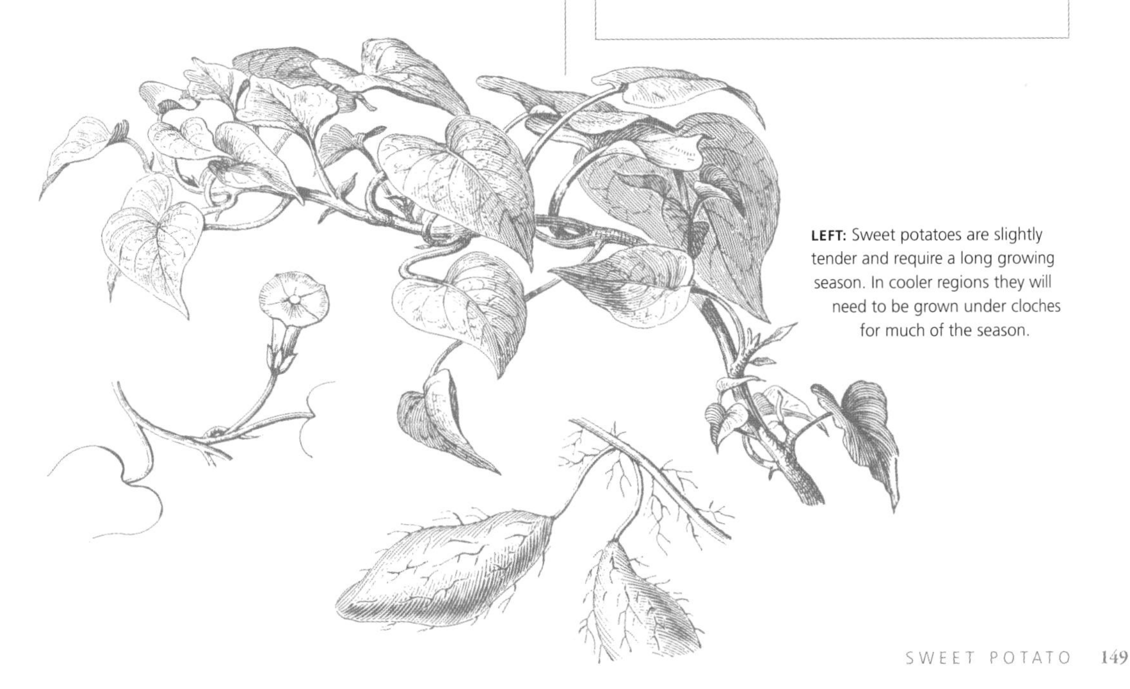

LEFT: Sweet potatoes are slightly tender and require a long growing season. In cooler regions they will need to be grown under cloches for much of the season.

Tasting notes

Sweet potato and beetroot

The colour and flavour combinations of deep red beetroot and mellow orange sweet potato make them a delicious accompaniment to rich meats, particularly duck or venison.

Preparation time: 10 minutes
Cooking time: 2 hours
Serves: 10 people (as a side dish)

- 4 beetroots
- 6 sweet potatoes
- 3 tbsp olive oil
- Salt and pepper, to taste

Pre-heat a conventional oven to 190°C (375°F / gas mark 5 / fan 170°C).

Cover beetroot with foil and place in the oven until tender – about 1–1½ hours.

Once cooled, peel the beetroot and cut into square chunks.

Cut the sweet potato into similar-size chunks.

Mix together, drizzle with olive oil and season.

Place on a baking tray and into the oven; cook until tender, approximately 40–60 minutes.

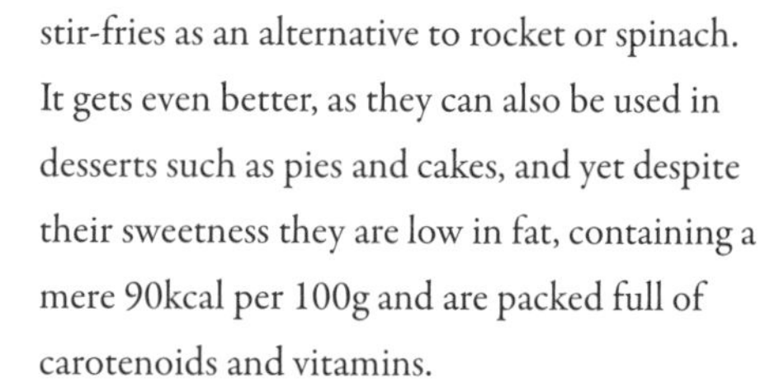

stir-fries as an alternative to rocket or spinach. It gets even better, as they can also be used in desserts such as pies and cakes, and yet despite their sweetness they are low in fat, containing a mere 90kcal per 100g and are packed full of carotenoids and vitamins.

Most sweet potatoes are grown from slips, which are leafy cuttings without roots. They can be ordered through seed companies, who will send them in the post. The slips may look a bit withered when they arrive, so will need to be revitalized for 24 hours by placing them immediately into a jar filled with a little water to cover their bases. Then pot them on into multipurpose compost and keep them on the windowsill for a few weeks before planting them out after the risk of frost is over.

The variety 'Beauregard Improved' has stunning bright orange flesh and one of the sweetest flavours. Other varieties worth trying include 'Georgia Jet' and 'O'Henry'.

If you can resist cooking the entire crop after harvesting, save some tubers in pots of moist compost in the potting shed to use as next year's stock. Move them into the glasshouse or sunny window ledge in spring; once they start producing shoots, these can be cut and planted up as slips for the new season.

LEFT: These sweet potato tubers are packed full of healthy goodness and can be used in a wide range of dishes, being suitable for boiling, grating, baking and making into chips.

Lettuce

Lactuca sativa

Common name: Lettuce

Type: Annual

Climate: Tender, frost-free winter

Size: 25cm

Origin: Europe and Mediterranean

History: Lettuce varieties appear as far back as 2700 BC, depicted on Egyptian tomb paintings. It is known to have been eaten by the ancient Greeks and the Romans cultivated several varieties including cos and butterheads. The first lettuce was introduced into England in the 16th century.

Cultivation: Seeds can be sown from early spring onwards in shallow drills. They are also ideal candidates for growing in containers, hanging baskets and window boxes due to their short root systems. They will tolerate some shade but it is important they are kept well watered as they need the moisture to make their quick growth. Some varieties are ready to harvest just a few weeks after sowing.

Storage: Loose-leaved lettuces do not last for much more than a day after harvesting before they start wilting and should only be picked when they are to be used that day. Heart-forming lettuces such as iceberg will keep for a few days in the fridge – place in a damp plastic bag to revive limp leaves. Sow little and often so that all the plants are not ready for picking at the same time as they can not be frozen or preserved.

Preparation: To prepare, trim the base stalk and remove any damaged outer leaves. Separate the remaining leaves and wash in a bowl of cold water. Dry thoroughly and tear the leaves into pieces. Lettuce is normally served raw in salads.

BELOW: Lettuce leaves are so easy to grow and can be used with almost any meal as either garnish, a side dish or as the main ingredient in many different types of salads.

Lettuce has been a summer favourite for centuries. It was commonly grown during Roman times, but dates back even further. The original lettuces would have been more bitter than the mellower and sweeter flavours we have become accustomed to now. In the past the leaves would have required blanching to make them palatable, whereas nowadays there is a huge range of crisp and crunchy succulent leaves for people to enjoy. It is one of the most commonly used vegetables, possibly because it does not need cooking – making it an ultimate fast food, but also because its mildness does not clash with other flavours. It is ideal as a sandwich or burger filling, for garnishing just about any meal from around the world.

Lettuces are not only beautiful when used as garnish in the kitchen; they also add a decorative quality to kitchen gardens and potagers. The range of leaf textures, from frilly open-leaved varieties to dense hearting types, makes them a must for anyone trying to create beauty in the vegetable plot. Colours vary enormously too from deep green to bright reds, meaning they are often used to create patterns in vegetable beds in much the same way as bedding plants. There are lots of different varieties available, some of which form a centre or heart, and others that are heartless.

Nutrition

Fresh leaves are an excellent source of vitamins A, C and K and beta carotenes. These compounds have antioxidant properties. vitamin A is required for maintaining healthy skin and vision and vitamin K for healthy bone function.

ABOVE: Lettuce can be prone to bolting if sown too early and placed under stress early in the season. This causes it to flower prematurely and run to seed, resulting in bitter-tasting leaves.

Lettuces are fairly versatile in the garden; they can tolerate a range of different soils and actually prefer some shade during the heat of the day. The soil should ideally be moisture retentive to ensure the lettuces can form their succulent leaves, so plenty of organic matter should be dug in prior to planting. Because of their fast growing habit and the fact that they do not take up much room, they can be planted amongst other crops and quickly harvested before slower plants have had a chance to expand.

The simplest way to grow them is simply in 1cm deep drills about 30cm apart; thin them out after germination to between 15 to 30cm depending on variety. Because lettuces do not keep for long, it is better to sow little and often throughout summer to avoid them all maturing at once. Because of their shallow rooting habit, they can also be grown in window boxes and containers in the same way.

For earlier crops, they can be sown indoors and planted outdoors under cloches. The season can also be extended into autumn by picking winter varieties such as 'Artic King', 'Rouge d'Hiver' and 'Density' (syn. 'Winter Density') and planted under cover.

Heart-forming lettuces are ready for harvesting once they feel plump, whereas the loose-leaved types can be picked when small and harvested with scissors or picked later as a whole plant. Harvest the plants before they get too old as they quickly deteriorate, wilt and provide a nourishing meal for slugs and snails.

BELOW: There are lots of different types of lettuce to try and they come in range of different colours and textures, making them useful plants for creating attractive patterns in the vegetable garden.

Tasting notes

Lettuce varieties to try

There are lots of different types of lettuce leaves worth trying. Essentially a summer crop, it is possible to harvest almost all year round by carefully selecting early and late varieties and growing them under cloches.

Butterheads	Fast growing and an open habit, with tender leaves but a soft succulent heart. Varieties include 'Artic King' 'Clarion', 'Diana', 'Roxy' and 'Sangria'.
Cos	Sometimes called romaine lettuces, these have an upright habit and are elongated. There are smaller semi-cos types too. Ones to give a go are 'Little Gem', 'Cosmic' 'Tan Tan', 'Corsair' and 'Little Leprechaun'.
Crispheads	As the name suggests, they have crisp succulent heads with a firm heart, the most commonly known lettuce being 'Iceberg'. They are less prone to bolting. Others to try are 'Lakeland', 'Sioux', 'Robinson' and 'Minigreen'.
Loose leaf	These varieties do not form a heart and are often treated as cut-and-come-again types which are simply harvested with scissors and left to resprout. Types include 'Red Salad Bowl', 'Salad Bowl', 'Frillice', Lollo Rossa' and 'Nika'.

CROP ROTATION

Long term planning is essential for the success of a vegetable garden. It is not just about thinking about the year ahead; it is considering what will be grown in each section of the garden for the next three or four years. Crop rotation is a popular system used by gardeners to plan what is going to be grown where. Vegetables are divided up into family groups and are then moved from one area of the garden to another each year.

There are no hard and fast rules, but the groups are usually divided as follows:

ABOVE: Peas and beans are part of the legume family. They fix nitrogen from the air and leave it in the soil after they are removed, creating a fertile bed for the following group of vegetables, the brassicas.

Group one:

ROOT CROPS AND POTATOES

- Potatoes
- Root crops (such as carrots, parsnips, beetroot, radishes)

Group two:

THE LEGUMES

- Peas, including mange tout and sugar snaps
- Beans including French beans, broad beans, runner beans

Group three:

BRASSICA FAMILY

- Kale
- Brussels sprouts
- Cabbages
- Broccoli
- Cauliflower
- Kohlrabi

Other vegetables such as lettuce, sweetcorn, squashes and pumpkins are usually just slotted in among the beds. Tomatoes are often planted in the same rotation as potatoes as they are from the same family. Perennial crops such as rhubarb, globe and Jerusalem artichokes and asparagus are not included in the rotation process as they stay in the same place from year to year. Nor does it include fruit trees and bushes, but it is important to take them into account when planning the vegetable beds.

THREE- AND FOUR-YEAR ROTATIONS

Most people use a three-year crop rotation plan. Occasionally a four-year plan is used, and this would involve splitting out potatoes from the root crops.

The rotation order in one bed for a three-year plan would be:

Year 1: Potatoes/tomatoes and root crops
Year 2: Legume family
Year 3: Cabbage family

The rotation order in one bed for a four-year plan would be:

Year 1: Potatoes
Year 2: Root crops
Year 3: Legume family
Year 4: Cabbage family

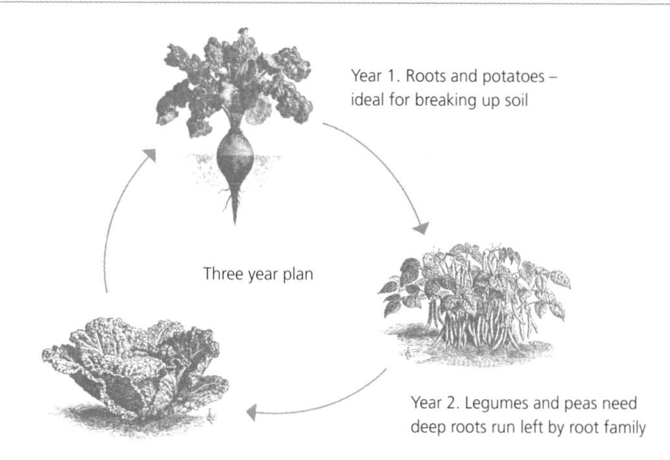

ABOVE: Most people operate a crop rotation with three groups of plants.

BENEFITS OF CROP ROTATION

By moving the crops into a different growing space each year, it prevents the build-up of soil-borne pests and diseases. Many of the problems in the soil are specific to one group of plants, such as club root affecting brassicas, or onion white rot affecting the onion family. Growing a different type of crop in the soil each year avoids on-going problems. In three or four years' time when the group of plants returns to the same soil, hopefully you will be one step ahead of the pests and diseases.

Crop rotation also assists with the fertility of the soil. Moving crops around avoids the constant depletion of the same nutrients. Different plant groups have specific requirements and can benefit from being planted following another crop. For example, the pea and bean family require a deep root run and therefore benefit from the space vacated by the root family such as carrots and potatoes. Likewise brassicas, which need a fertile soil, do better when planted after the pea and bean family as the soil is enriched with nitrogen left by the legumes' root system.

> 'Seeing a fruitful and pleasant Garden, can not be had without good skill and diligent minde of the Gardener on matter of the ground.'
>
> **Thomas Hyll, *The Gardener's Labyrinth*, (1577)**

Asparagus pea

Lotus tetragonolobus

Common name: Asparagus pea, winged lotus

Type: Annual

Climate: Tender, frost-free winter

Size: 40cm

Origin: Europe, Mediterranean

History: Little archaeological evidence has been recorded on the early history of the plant or its uses, but it is mentioned in many Renaissance herbals. There are records of its cultivation in Sicily from the mid 16th century.

Cultivation: Either sow them directly into fertile soil in full sun or in modules. Wait until the risk of spring frosts is over. Keep them well watered and weed free and harvest the pods regularly to encourage more pods to develop.

Storage: The pods will last a few days in the fridge, but the best way to preserve their summer flavour is to cook them in soups and freeze them, use them instead of beans in pickles and preserves or make a type of piccalilli relish from them.

Preparation: Only the very young seedpods, when less than 2cm long, can be used in the kitchen either raw or cooked. They can be added to salads, lightly steamed as a vegetable and served with melted butter, or added to soups and stews.

Capture the essence of asparagus in a pea pod produced by a beautiful-looking plant that really will put the wow factor into the vegetable patch. It is a small shrubby-looking plant that produces stunning red-maroon flowers with dark centres that look a bit like sweet peas. The blooms are followed by strange, quirky-looking winged seed pods, which are the parts of the plant that are cooked. Popular but expensive to buy in Asia, and practically impossible to purchase closer to home, the only way to get hold of these delicious but curious-looking seed pods is to grow your own.

The benefits of growing this plant do not end there. As it is a legume or member of the pea and bean family it means the roots of the plant fix nitrogen from the air. This is great for the environment as it means less need to fertilize and therefore nutrient hungry plants such as members of the cabbage family can be planted in this fertile-rich soil the following year. It is a much easier plant to grow than standard asparagus as it does not mean having to wait a couple of years for it to crop, and does not take anywhere near the same amount of space as the perennial plants. So if you want to flavour your dishes with asparagus but do not have the room or patience, then asparagus pea is an absolute must.

The simplest way to cook the pods is to lightly steam them, adding butter and salt, although they can be enjoyed raw. However, they can be used in a whole range of dishes and used as a substitute for pea or asparagus in recipes. They should be sliced and can then be sautéed, grilled or fried and are delicious when used as a starter by dipping them into rich and spicy sauces such as sweet and sour or mango chutney. Their crunchy texture also goes well with fish, prawns and chicken dishes and frying them up with a slice of goat's cheese, red onion, garlic and a handful of herbs brings out the flavour. A delicate soup can also be made from these pods.

Seeds should be sown in 7.5cm pots or modules in potting compost under glass from mid to late spring and left to germinate on a sunny windowsill or glasshouse. After a few weeks they should be hardened off in a cold frame or even the porch for

LEFT: Considered to be a bit of curiosity rather than a staple vegetable in the garden, asparagus peas are very easy to grow and in addition produce attractive flowers.

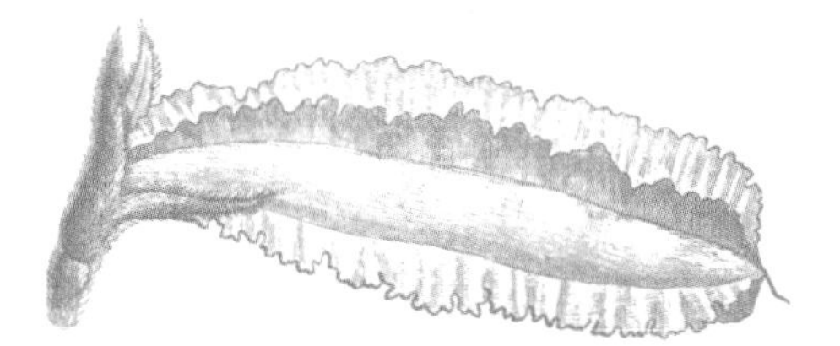

ABOVE: The edible part is the winged seed pod, which is very popular in Asian cuisine. Harvest the pods when they are about 2cm long. Any longer and they become unpalatable.

about 10 days before planting them out after the risk of spring frosts is over. Plant them in a sunny, sheltered spot at 30cm apart. They prefer light fertile soil that is well-drained and a generous helping of garden compost to be added to the soil to get them started.

Alternatively, they can be sown directly outdoors after the risk of spring frosts is over. Placing a glass or plastic cloche for the first few weeks after sowing will get the plant off to a quick start. Sow them every 30cm and give them a good watering afterwards. Depending on their vigour they may need twiggy supports but it is not usually necessary. Plants should be kept weed free although their trailing habit does form a useful mat that helps suppress encroaching weeds.

Due to their attractive trailing habit they are also suitable for growing in a hanging basket or window box, but it will need regular watering.

Harvest the pods when they reach about 2cm. Do not allow them to get much longer because they lose their texture and become stringy and unpalatable. Like sweet peas or runner beans it is important to keep picking the pods because otherwise they stop cropping. When they have finished cropping they should be chopped down at ground level and the nitrogen-enriching roots later dug over into the soil.

Watercress

Nasturtium officinale

Common name: Watercress

Type: Perennial

Climate: Very hardy, very cold winter

Size: 20cm

Origin: Europe

History: Watercress has a long history dating back to the Greeks, Persians and Romans and is the most ancient of green vegetables known to man. It was not until the early 19th century in England that watercress was cultivated on a large scale.

Cultivation: Grow in moist fertile soil or in a container such as a washing-up bowl. It can be grown in compost in a container or trough sitting on a saucer filled with water. Sow seeds in situ during spring or summer, or in trays indoors during winter.

Storage: Leaves do not last for long so pick as required. Alternatively, make into soup and freeze.

Preparation: Wash well and trim off the tough stalks before use.

ABOVE RIGHT: Watercress is very easy to grow and has a strong, peppery aroma. Try to keep plants contained if planting directly into damp ground as they will rapidly spread.

You do not need a vast riverbed to enjoy the peppery flavours and health benefits of watercress. All that is needed is a shady corner of the garden and a container to hold the water in. Suddenly the world of home-grown watercress soup is available to anyone with the smallest of gardens. This peppery, leafy herb with its straggly habit is a fantastic accompaniment to strong-flavoured meats such as boar and venison, but works equally well in poultry dishes. It is usually served raw as garnish or healthy sandwich fillers. Watercress combined with orange segments and a splash of olive oil is a classic side dish or salad.

Watercress must be kept permanently damp for it to thrive but it does not have to be permanently submerged in water, and can instead be planted in boggy soil in the garden. However, in the right conditions, it will quickly spread and take over, so think carefully before doing this. The easiest method of obtaining home-grown watercress is to grow it directly in a container filled with compost. It can be anything from a washing-up bowl to an old sink. If the container has drainage holes then it should be placed on a saucer filled with water so that it can absorb moisture. Seeds should be sown on the surface of damp compost during spring or summer in a container. Keep the container or saucer regularly topped up with water. Occasionally pour water through the container to keep the water fresh and regularly change in the saucer to prevent it going stagnant. It is tough and will supply leaves throughout the year, staying green until about –2 or –3°C .

Growing without a garden

If you do not have a garden and still crave some of these peppery leaves then do not despair. They can be grown on a window sill and regularly harvested as baby leaves throughout the year. Lightly scatter them on a seed tray containing compost, cover with a clear plastic bag and leave them to germinate. Keep the tray well supplied with water and regularly topped up. Regularly harvest leaves when they are big enough to handle and add to salads or sandwiches.

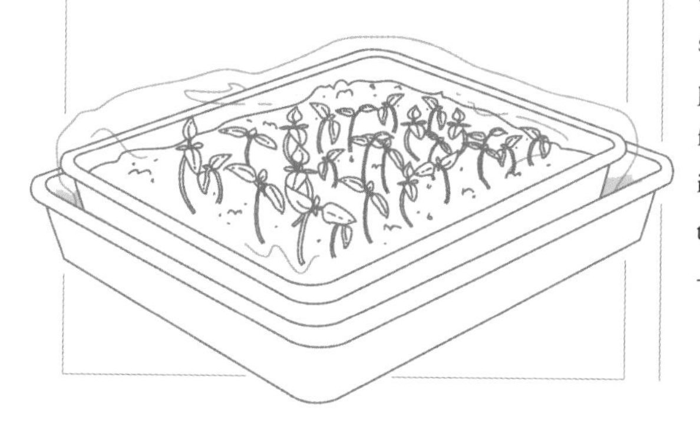

Past roots

Watercress was a popular vegetable in London during Victorian times along the riverbanks as there was such a shortage of nutritional greens throughout the year, particularly winter. Due to its popularity it is thought it prevented outbreaks of scurvy.

Throughout history watercress has been highly valued for its medicinal uses. In fact, a famous Persian chronicler advised Persians to feed watercress to their children to improve bodily growth. It was also given to the Roman soldiers as it appeared to improve their health and made them stronger. When Hippocrates founded the first hospital on the Island of Kos around 400 BC, he grew wild watercress in the natural springs and used it to treat blood disorders.

Hippocrates
(c. 460–370 BC)

Parsnip

Pastinaca sativa

Common name: Parsnip

Type: Annual

Climate: Hardy, cold winter

Size: 25cm

Origin: Eastern Mediterranean

History: Archaeological evidence has shown parsnips to have been around since prehistoric times. Parsnips grew wild in Europe and were considered a luxury item for the aristocracy in ancient Rome. The Emperor Tiberius accepted part of the tribute payable to Rome by Germany in the form of parsnips.

Cultivation: Parsnips require a long growing season so seeds should be sown directly outdoors from early to mid spring. They require a fertile, well-drained soil. Sow rows 30cm apart and thin so that they are 15cm apart within the rows. Keep them regularly watered as they tend to split if it rains heavily after a dry period. Harvest from autumn onwards.

Storage: Parsnips can remain in the ground throughout winter, but in very cold areas it might be difficult to dig them up if the soil freezes over. They should therefore be lifted in autumn and buried in trays filled with sand or compost and kept in a cool, dry place such as a garage.

Preparation: Scrub well, trim the top and root ends and peel thinly. Either leave young parsnips whole or slice large parsnips into quarters and remove the central core. Boil in salted water, steam for 15–20 minutes or roast around a joint of meat after par boiling.

Not offering much above ground in terms of beauty in the garden, this winter root crop does all its talking below the surface as it is the aromatic and sweetly flavoured tap root that gardeners and chefs love. It is one of the classic vegetables to accompany a traditional roast meal when glazed with honey and roasted, but there is so much more to this old favourite. They can be made into fancy herby chips, or thinly sliced, drizzled with chilli oil and baked in the oven to make crunchy crisps. They can also be mashed and mixed with maple syrup or grated into a rosti and fried. They can also be curried or added to traditional hotpots, casseroles, soups and winter stews and combined with other winter vegetables such as leeks, turnips and swede.

There are lots of varieties to chose from but it is worth choosing one that has some canker resistance as it can be heart breaking to dig up the roots in autumn and discover they are all split and rotting. Resistant varieties include 'Gladiator', 'Albion' 'Palace' and 'Archer'. One of the best flavoured varieties is called 'The Student', which is an old heritage type dating back to 1800s and also has canker resistance. For those who enjoy a

RIGHT: The chromolithograph plate from the *Album Benary* of 1876 shows a range of parsnip and other root varieties.

gourmet treat then try the variety 'Arrow', which has narrow shoulders, meaning they can be grown closer together at high density, which will provide the sweet 'baby' parsnips that are so popular in good-quality restaurants.

Often it is advised that seed can be sown as early as February because parsnips do need a long growing season, but unless it is an exceptionally mild spring and the soil is not cold and heavy, the seed is likely to languish and eventually rot in the soil. It is better to wait until late March or early April. Seed should be sown in shallow drills 30cm apart. Sow three seeds at stations every 15–20cm; gently rake back over the soil and water. Thin the seeds out to one seed when they are about 5cm tall. As they can be slow to germinate, it is possible to sow a row of radishes in the row too as this helps mark out the row and they can be quickly harvested without affecting the emerging parsnip crops.

Like carrot seed, avoid using old stock as it quickly goes stale. Instead buy fresh seed each year to ensure reliable germination and a bumper crop. Avoid freshly manured soil as this can cause the parsnips to fork. Instead, if possible use fertile, free-draining ground that has been manured in a previous year. Stony soil should be avoided as well as this prevents the root from developing fully. One technique to encourage a deeply rooted parsnip and to avoid the problem of stones is to use a crow bar at each sowing station to push it down deeply into the soil; then fill it up with compost. Keep the plant well watered throughout summer and keep the area weed free. In very cold areas it may be necessary to protect the tops of the parsnips with straw, but they are generally very hardy.

Parsnips are ready for harvesting from autumn onwards, although it is possible to lift them earlier in the year for smaller roots. Most connoisseurs insist on waiting until the first autumn frosts before digging them up as they are supposed to taste sweeter. Use a fork to loosen the ground, taking care not to damage the root, and then lift them out by hand.

LEFT: Parsnips should be directly sown in the ground early on in the season. They are slow to germinate but will provide a crop from late summer and throughout winter.

'The parsnip is a very nourishing and valuable vegetable for winter food, and is highly esteemed by those who like vegetables of a sweetish, agreeable flavour.'

T. W. Sanders, *Kitchen Garden and Allotment*, (c.1920)

Runner bean

Phaseolus coccineus

Common name: Runner bean, scarlet runner, string bean or stick bean

Type: Annual

Climate: Tender, frost-free winter

Size: 2m

Origin: Central America

History: They were introduced to Europe from Mexico in the mid 17th century. However, runner beans have been known as a food crop for well over 2,000 years. They were introduced to Britain in the 17th century by the plant hunter John Tradescant (gardener to King Charles I) and were grown as a decorative plant before being used as a food in Britain. Today they are eaten in each of the five continents.

Cultivation: Beans should be sown in full sun in fertile soil. Seeds should be sown from mid to late spring under cover and planted out after the risk of frosts has passed. Keep them well watered and harvest when the beans inside the pod have started to swell and have reached about 15cm long, depending on variety.

Storage: Beans do not last for long after they are picked but they can be frozen or preserved in relishes.

Preparation: Cut off the ends and remove the strings from the sides by running a sharp knife down each side of the bean. Cut diagonally into 2.5cm (1in) lengths or slice into lengths. Steam or cook in salted boiling water for about 10 minutes.

BELOW: Runner beans can also be grown as an ornamental annual climber. Because of their attractive flowers and rapid growth they make a decorative and edible screen.

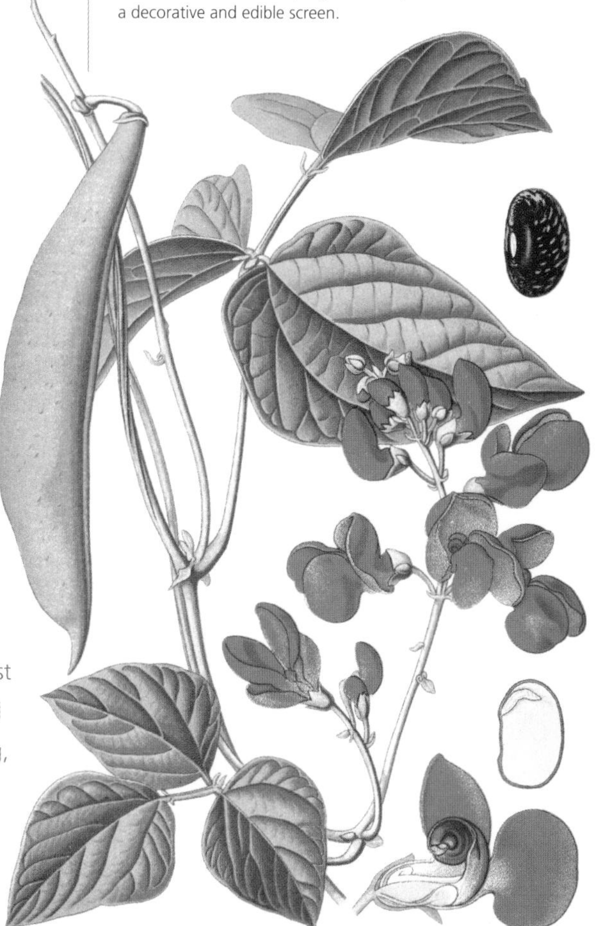

LEFT: Runner beans have a climbing habit, using tendrils to scramble upwards. They will need a structure such as hazel sticks for support.

A rustic wigwam smothered with brightly coloured flowers and strings of tender runner beans is a quintessential kitchen garden image. It is not hard to see why the first plants introduced to Europe in the 16th century were grown more for their ornamental value, before people realized how delicious the seed pods were. Runner beans not only add height, colour and structure to the otherwise flat-looking vegetable plot but home-grown varieties taste so much better than the ones bought in shops. There is a huge variety of dishes that runner beans can be used for in the kitchen. They should be eaten cooked, usually steamed or boiled, and thinly sliced to maximize their flavour. Their flowers are also edible and commonly added to bean salads or used as garnish.

One of the reasons for their popularity on the vegetable plot is for the huge gluts that they provide during summer. Not only will they keep cropping if regularly picked, but any excess can be frozen, meaning there is hardly any wastage and nothing needs to get slung on the compost heap. Despite their height, they can also be grown on small balconies and patios as they can be planted in deep containers that are at least 50cm in diameter and have plenty of drainage holes. Containers should be filled with good-quality compost. The plants will need watering daily and feeding once a week once they start to flower.

Runner beans require a sunny, well-drained plot with plenty of organic matter added to the soil to improve its ability to retain moisture and to add some nutrients. A runner bean trench can be created the autumn before planting, which involves digging out a 30cm trench where the beans are to be planted later. It should be filled up over winter with kitchen waste and other organic matter, gradually letting it rot down so that when it comes to planting out the runner beans they have a beautiful, fertile soil to grow in. Runner beans should be grown in a sheltered spot, partly to prevent the tall climbing structures being blown over.

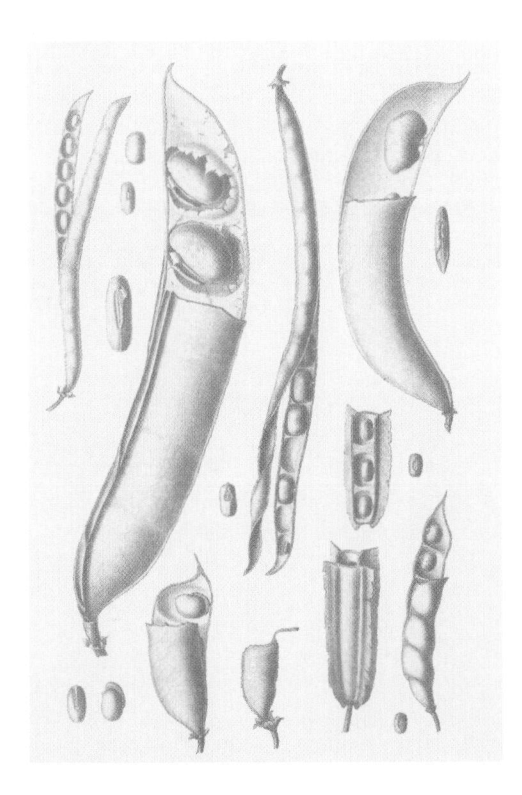

LEFT: Left to mature, the pretty seeds can be dried and stored for sowing the following year.

Creating a climbing structure

Runner beans are climbing plants and use their tendrils to scramble up structures. To make it easier for picking, they are usually trained up tepee or wigwam structures made from bamboo canes or hazel sticks. They are usually about 2m tall although most runner beans will climb much higher if allowed, but this makes picking harder.

To make a tepee 2.4m canes should be pushed into the ground in a circle, leaving about 25cm between each one. The tops of the canes should be pulled tightly together and lashed with gardening twine. Alternatively, they can be trained on a pair of parallel rows of canes, whereby they are joined by a horizontal cane at the top running parallel with the row. Rows need to be 60cm apart with canes spaced 20cm apart.

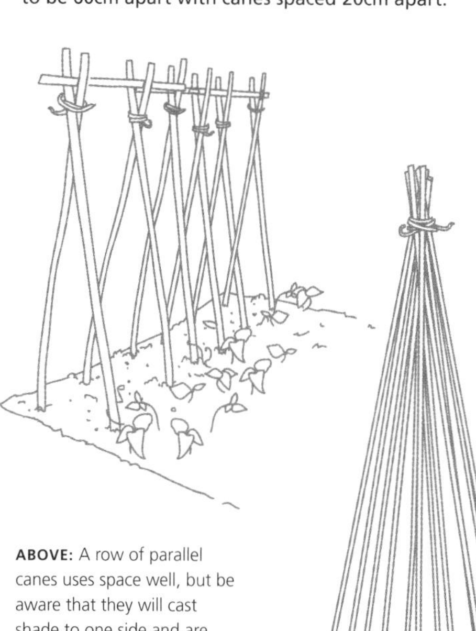

ABOVE: A row of parallel canes uses space well, but be aware that they will cast shade to one side and are prone to being blown over on exposed sites.

RIGHT: Left to climb, runner beans will keep going and may pull over the entire support. Stop them when they reach the top.

Seeds can either be sown indoors in mid-spring or outside after the risk of the frosts. The advantage of growing them indoors to start with is that it gives them a head start for the season ahead. Sow one seed per 7.5cm pot at 3cm deep in a multipurpose compost. Water the seeds and leave them to germinate on a sunny window ledge or in the glasshouse. Before planting them directly outside they should be hardened off in a cold frame for a few days to gently acclimatize them to the outside weather. Plant them at the base of each bamboo support and water them well. Pay particular attention to watering once the flowers start to form.

Alternatively, the seeds can be sown in early summer by using a dibber to push two seeds 3cm deep at each cane support. Once germination has taken place the weaker one of the two seedlings should be removed. The plants will need regular watering to help them maintain their vigorous, climbing growth habit. On dry soil it is also beneficial to mulch around the root system with garden compost to keep the root system moist. Harvest the beans regularly and do not let them get too long as they become stringy and tough.

LEFT: Wigwams make solid supports but slightly less space efficient as they present less of a surface area for the beans.

GARDEN TOOLS

Man has toiled the soil for thousands of years, yet many of the tools remain the same today as back in the old days. Having the right tool for the right job is essential, as it increases efficiency and can save hours of backbreaking work. When purchasing tools it is important that they are suitable for your height and strength. Otherwise you run the risk of over-extending and injuring yourself. If good-quality tools are looked after properly, they should last a lifetime.

Soil dibbers

String lines

Below are some of the essential tools required for creating and maintaining a vegetable garden.

DIBBER

This slender hand tool is pushed into the ground to make holes for seed sowing. It can either be bought, or simply made from off-cuts from bamboo canes and sticks.

STRING LINE

Vegetable gardeners love their straight lines. A string line is used prior to digging out a trench, making a seed drill or making holes with a dibber. It can also be used for marking out paths or rasied beds.

SPADE

Probably the most essential tool for vegetable gardening. Use it for digging over the soil but be aware – it slices through perennial weeds roots, which increases the problem.

Sheds and tool care

Garden tools should be stored away under cover when they are not being used, ideally in a shed as it provides plenty of space. Sheds can also double-up as a potting area or simply somewhere to shelter from the rain and have a cup of tea. It should have lots of shelves and racking so that tools, pots, bags of compost and other garden implements can be secured safely.

Wipe mud and dirt off all tools before storing them away. Metallic material will benefit from being wiped down with an oily rag, while wooden handles can be occasionally rubbed with linseed oil. Regularly sharpen implements used to cut, such as secateurs, loppers and hoes.

SHOVEL

An invaluable tool used for moving loose materials such as garden compost, manure and wood chippings around the vegetable plot. When purchasing one, check the weight and balance so that you do not damage your back.

FORK

Used for digging, breaking up the ground, particularly on heavy soils, and reducing compaction. The back of the fork can be used for breaking up clods of earth.

RAKE – LANDSCAPE

Landscape rakes are regularly used on the vegetable plot for levelling out the soil after digging and spreading compost and manure. They have a larger head than traditional rakes, making levelling easier.

WHEELBARROW

Where would gardeners be without this essential bit of kit for moving plants, compost and garden waste around the garden? Paths should be wide enough in the kitchen garden to accomodate a wheel barrow.

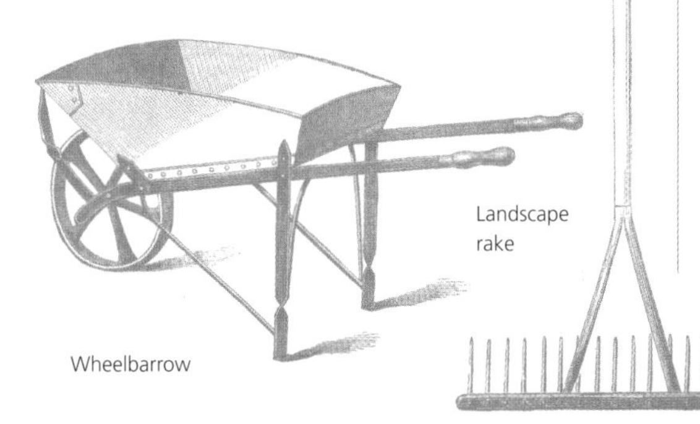

Landscape rake

Wheelbarrow

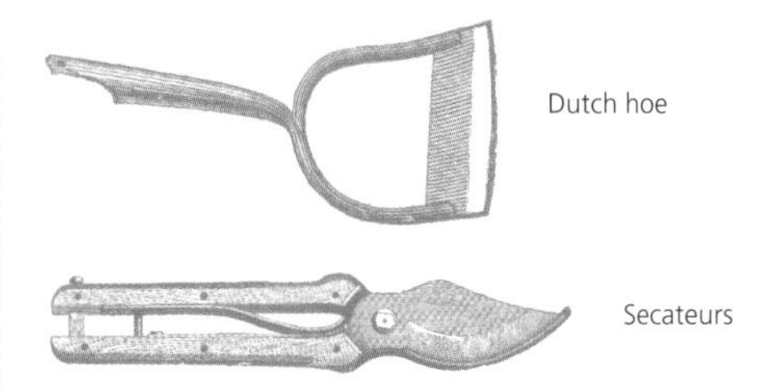

Dutch hoe

Secateurs

DUTCH HOE

This essential tool is used for removing annual weeds between rows of vegetables. It cuts on both the pushing and pulling motion and the long handle avoids backbreaking weeding.

PRONGED CULTIVATORS

These strange-looking implements can be used to scratch at the surface of the soil, breaking up compaction and allowing water to penetrate beneath the surface.

SOIL SIEVE

If you make your own compost then a soil sieve is useful for removing large un-composted material. It can also be used for removing stones in the soil.

SECATEURS

These are used to cut back garden foliage and to harvest vegetables. Bypass-type secateurs are more expensive but give a better cut than the anvil types. A pocket knife is also useful.

WATERING CAN

All seeds and seedlings should be watered after planting. A rose can be fitted to the watering can's nozzle so that the water does not wash away recently sown seeds or emerging seedlings.

French bean

Phaseolus vulgaris

Common name: French bean, common bean, snap bean, green bean, haricot bean (dried)

Type: Annual

Climate: Tender, frost-free winter

Size: 2m

Origin: South America

History: These beans have been grown as a crop for thousands of years in South America. Archaeologists working in Peru have dated bean remains to about 5000 BC. They were first introduced to Europe in the 16th century by Spanish and Portuguese explorers.

Cultivation: French beans are frost tender so should only be grown during the summer months. Train climbing beans up tepees or climbing structures. They require lots of moisture, sunshine and organic matter for them to grow successfully.

Nutrition

French beans are a good source of protein, thiamin, riboflavin, niacin, vitamin B6, calcium and iron, and a very high in fibre. They also contain good levels of vitamins A, C and K.

Tasting notes

Attractive varieties to try

Although there is not much difference in flavour between French beans, their range of colour adds a visual vibrancy to any dishes.

'Borlotto Lingua di Fuoco'	A dwarf variety with flat pods and speckled with red.
'Purple Teepee'	A dwarf variety with purple pods that go dark green when cooked.
'Cobra'	A climbing type with tender green pods but attractive purple flowers.
'Selma Zebra'	An heirloom climber with quirky looking green streaky pods.
'Golden Gate'	A climbing variety with impressive bright yellow pods.

Storage: French beans are best eaten fresh but will store in the fridge for a few days. They can be frozen but should be blanched first. If the pods are left to swell and dry these haricots can be stored in sterilized, airtight containers for months.

Preparation: Choose slim french beans that break with a crisp snap. The beans should be young and only need topping and tailing, but if they are a little coarse then they may need their stringy sides removed. They can then be steamed or cooked in boiling salted water for about 10 minutes.

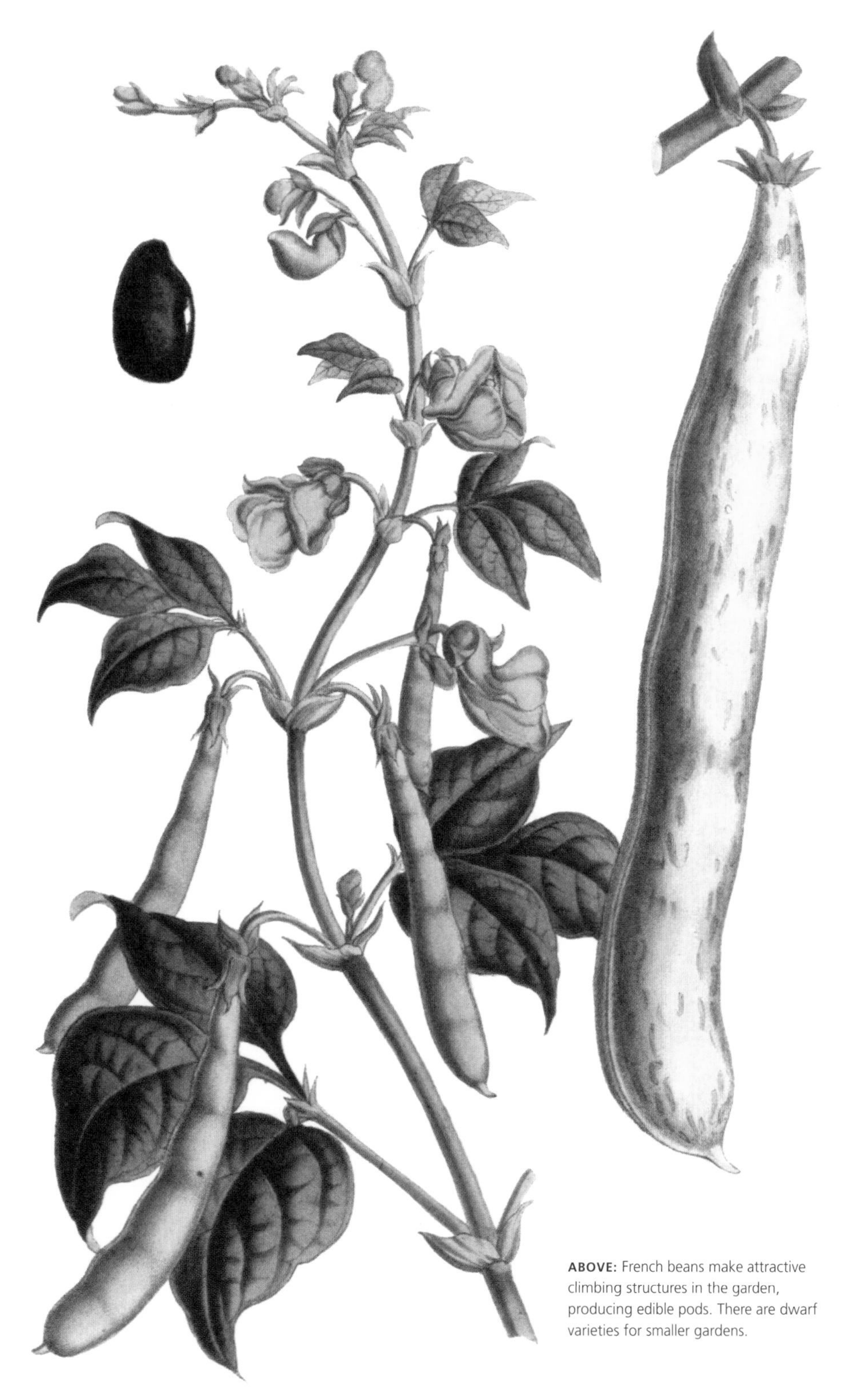

ABOVE: French beans make attractive climbing structures in the garden, producing edible pods. There are dwarf varieties for smaller gardens.

French beans have been grown for over 7,000 years and are more popular now than ever before. They are so easy to grow and yet are very often overlooked in summer by their closely related cousins, the runner bean. Yet for those who find runner beans too stringy, these are the answer as they are far less likely to go that way. They are usually grown for their pods and, unlike runner beans, these come in a range of bright colours, which is ironic really considering their other common name is 'green bean'. Look out for varieties coming in yellow, purple, cream and speckled. These brightly coloured pods make them a sight to rival any ornamental flower display and some gardeners grow the plants for that reason alone. One other advantage French beans have over runners is that they come into cropping earlier, sometimes as quickly as seven or eight weeks after sowing, making them useful for filling that summer bean gap in the harvest calendar after the broad beans have been harvested. There are dwarf or bush varieties too, only growing to about 50cm high, meaning they can easily be grown on a sunny patio or balcony without having to overstretch. The shorter versions are great in small vegetable patches where shading neighbouring plants can be a problem.

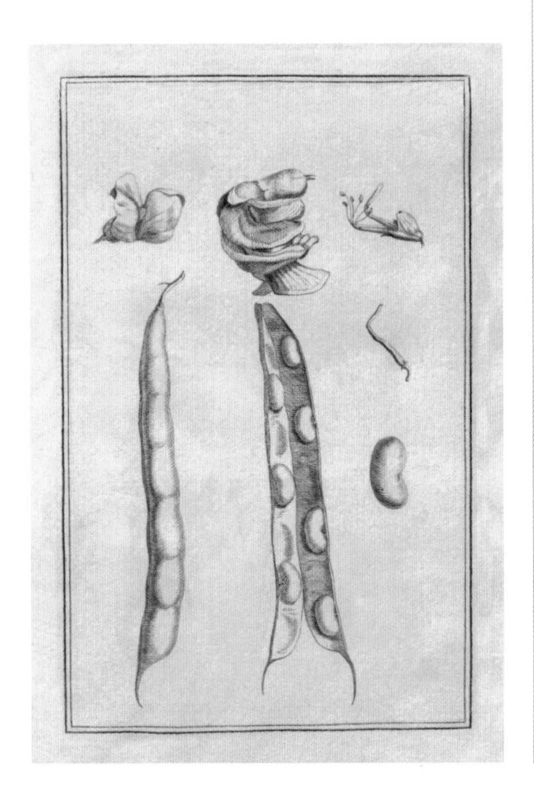

French beans are so easy and quick in the kitchen too. Simply pick them fresh from the garden, top and tail them and boil them in salt water for three or four minutes. Afterwards run them under cold water to prevent them overcooking or going soggy, slice and toss them into salads and add a dressing. Their flavour is versatile, meaning they will go with anything from a lemon-based vinaigrette to something stronger such as a soy sauce-based mix. Combine them with fried chillies, garlic and peppers and a pinch of flaked almonds for a delicious quick and healthy snack.

To produce haricot (sometimes called navy beans) the pods are left until the end of the season and then cut down and left to dry in a frost-free place such as a garage. Once they are 'rattle dry' they can be shelled and stored in airtight containers. They have a mild flavour, meaning they are a perfect accompaniment for bulking up stronger flavoured foods. They are commonly added to cassoulet and other slow-cooked dishes, but can equally be used in bean salads, soups or even in purées.

LEFT: French beans are usually harvested when reasonably young, but the beans inside the pod can be left to swell and then picked, dried and stored as haricot beans.

French beans like to be in full sun and require the same fertile and nutrient-rich conditions as runner beans. A 'runner bean trench' can be dug the autumn before planting to enrich the soil (see p.164). Alternatively dig in plenty of garden compost in spring. Seeds can be sown indoors in plastic pots in mid-spring and kept on a sunny window ledge or glasshouse for germination. They should not be planted out until after the last of the spring frosts. Acclimatize the plants to the outdoors by placing them in a porch or a cold frame for a few days before plunging them outside into the elements. Climbing varieties will need a structure to scramble up (see p.165), and they should be planted out at 20cm between each plant and 90cm between rows. The dwarf beans should be 8cm apart in rows 45cm apart.

Alternatively seeds can be sown directly into the soil. Sow two seeds per 4cm deep hole using a dibber at the spacing mentioned above; once they have germinated the weaker one should be removed, leaving the other to grow away strongly. The growing tip should find the climbing structure on its own and start to twist its way upwards, but it sometimes needs help to find the canes at the early stages of development. Once attached it should romp up those canes. Then, when it reaches the top of the structure, the growing tip should be pinched out, which will encourage it to channel its energy into producing beans.

Harvest the beans regularly. This can start as early as seven or eight weeks after sowing. If you want to produce haricot beans then leave them to mature and swell, and harvest them right at the end of the season.

Tasting Notes

French beans with garlic

French beans taste best when cooked immediately after harvesting, as this captures those fresh flavours that are quickly lost if stored in the fridge. For a milder flavour, substitute garlic for elephant garlic or fried red caramelized onions. This simple dish brings out the natural herbaceous flavours of the beans, and only takes minutes to prepare.

Preparation time: 5 minutes
Cooking time: 5 minutes
Serves: 4 people (as a side dish)

- 700g (25oz) French beans, ends removed
- 30g (1oz) unsalted butter
- 1 tbsp olive oil
- 2 garlic cloves, crushed
- Salt and pepper, to taste

Blanch the French beans in a saucepan of boiling salted water for just 1½ minutes.

Drain and immerse in a large bowl of ice water to stop the cooking. When they are cool, drain and set aside.

Heat the butter and olive oil in a very large pan over a medium heat and cook the garlic for 1–2 minutes, or until lightly browned.

Add the French beans, sprinkle with salt and pepper, and toss together.

Reheat the French beans and serve.

Pea

Pisum sativum

ABOVE: Peas are one of the most popular vegetables in the culinary world, but their young emerging tendrils and shoots can also be eaten as a delicacy with herbaceous flavours.

Common name: Pea, mangetout, sugarsnap pea

Type: Annual

Climate: Tender, frost-free winter

Size: 1.2m

Origin: Middle Asia

History: Peas are one of the oldest cultivated vegetables and were believed to have grown at least as far back as 7800 BC. Archaeological remains of Bronze Age villages in Switzerland contained early traces of peas dating back to 3000 BC. The Greeks and Romans were cultivating peas from about 500 to 400 BC.

Cultivation: Sow early varieties in late autumn and other types in spring. Sow them in flat-bottomed trenches in fertile, well-drained soil in full sun. Provide twiggy pea sticks cut from hazel or birch trees to support the plants.

Nutrition

Peas are very high in fibre and can help lower cholesterol. They provide an abundance of nutrients, including iron in good levels. Peas are also rich in vitamin C, which helps maintain your immune system.

Storage: Peas must be one of the best vegetables for storing. They are without doubt best when eaten fresh from the plant, but their sweetness can be captured when frozen. They can also be dried although this is less common these days.

Preparation: Very full pods may have tough peas inside. Remove the peas from their pods and discard any that are blemished or discoloured. Wash under cold running water. Boil in lightly salted water, with a sprig of mint if desired, for about 10–15 minutes. Drain well and add butter. If steaming then do so for about 3–5 minutes.

Peas are one of the oldest cultivated vegetables in existence and they are just as much loved today. Fresh peas in a pod straight from the garden provide a flavour that is impossible to replicate from shop-bought ones and just for that reason alone they should be given pride of place in the vegetable patch. Many kitchen gardeners even struggle to get their pods as far as the kitchen before giving in to the heavenly delight of devouring the seeds from a freshly opened pod seconds after picking them. The secret as to why shop-bought ones just do not taste anywhere nearly as good is because as soon as they are picked the sweet sugars immediately start to transform into starch. Therefore, for lovers of the finest fresh peas, only home-grown ones will do.

Despite their delicious fresh flavours they are included in countless recipes from around the world, ranging from pea soups to the classic fish and chip accompaniment of mushy peas with mint. They are used in dishes in Europe, the Americas and particularly Asian cuisine such as classic Indian dishes including Mutter Paneer or Aloo Mutter (*Mutter* meaning 'pea').

For those who like a real gourmet treat the fresh shoots and tendrils can also be eaten early in the season and have a spinach-like flavour. Despite being a common sight at the markets in Asia they are practically impossible to get hold of further afield. To maximize their flavours the shoots

Sowing peas in lengths of guttering

Sowing into cold wet soil should be avoided as the seed can rot. If this is the case, then they can be sown in the glasshouse into short lengths of plastic guttering. Drill drainage holes in the bottom of the guttering and fill it up with general-purpose compost. Pop seeds in at a spacing of about 8cm and 3cm deep. Keep watered until the seeds germinate and then simply slide the soil and plants out into a shallow trench in the vegetable patch and water them in.

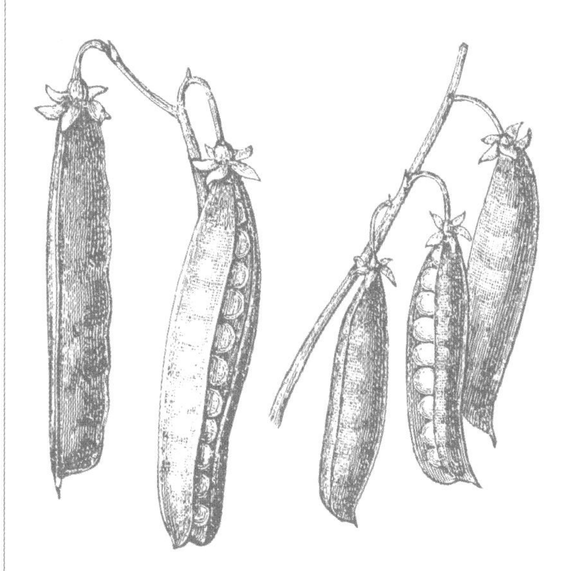

ABOVE: Most pea varieties need to be shelled from the pod before eating, but other varieties such as mangetout and sugarsnaps are eaten entirely, pod and all.

'The culinary pea is universally acknowledged to be the king of vegetables. It has been grown from time immemorial on the Continent and in this country, and it is regarded as one of the most delicious, as well as the most nutritious, of vegetables.'

T. W. Sanders, *Kitchen Garden and Allotment*, (c. 1920)

Tasting notes

Pea varieties to try

There are several different types of peas. Round peas, sometimes called smooth-seed types, are hardier and suitable for earlier sowings of seed. However, they do not taste as sweet at the less-hardy wrinkled types. The small, sweet-tasting peas that are so popular in the frozen aisles of the supermarket are known as petit pois and they can also be grown in the garden.

Round types	'Bountiful', 'Feltham First' 'Meteor' and 'Pilot'
Wrinkled	'Early Onward', 'Little Marvel' 'Hurst Green-shaft' and 'Onward'
Petit Pois	'Peawee' and 'Waverex'
Mangetout peas	'Delikata', 'Oregon Giant' and 'Snow Wind'
Sugarsnap	'Sugar Ann' and 'Sugar Bon'

The true gourmet fan will love mangetout 'Shiraz'. These unusual pods are purple with bicoloured flowers. Not only do they look great in both the garden and on the plate, but they are packed full of nutritional goodness including anthocyanin (antioxidants) pigments. They are best eaten raw, such as in a salad, but will retain some of their unique colour if steamed. If they are boiled they will lose their colour but still taste fantastic.

should be picked before they have a chance to open. There are lots of different types of peas to choose from, with some of the traditional types making room for some of the new kids on the veggie block, such as the sweet mangetout and sugarsnaps. These must be the ultimate healthy fast food, simply being picked from the plant and eaten immediately to maximize their crunchy, sweet and herbaceous flavours. Both are eaten whole, as the French name '*mange tout*' suggests.

Peas require an open but sheltered site in full sun. The soil should be thoroughly dug over before sowing and plenty of garden compost added. The early varieties can be sown in autumn and will overwinter but may need a cloche in cold areas. Alternatively, seed sowing should start in late winter or early spring and can continue through to early summer for some of the main crops.

The simplest way to sow the seed is to draw out a flat-bottomed trench using a draw hoe. The trench should be about 15cm wide and 5cm deep. Peas are sown in parallel rows either sides of the trench in a zig zag pattern. The soil should gently be raked back over the seeds and the area should be watered. Supports such as twiggy sticks or pea netting should be added when the plants have reached about 8cm high, taking care not to damage the plants.

Early peas have a speedy turnaround and are ready for harvesting about 11 weeks after sowing. The later varieties take a bit longer to mature. The podding types are ready for picking when the peas feel plump inside their pods. Picking should begin at the bottom of the plant, and work upwards as the season progresses.

Mangetout are picked when the pods reach about 7cm and should be caught before the seeds have swelled. Sugarsnaps should be about the same length but the pods should feel plump and the peas developed inside.

At the end of the season, cut down the plants to ground level and add to the compost heap. Chop up and dig the roots into the soil, though, as they are a valuable source of nitrogen.

TASTING NOTES

Tagliatelle with mint and pea pesto

A perfect combination of herbaceous fresh mint and pea flavours with the comforting background flavour of tagliatelle make this a wonderful light lunch or evening snack.

Preparation time: 5 minutes
Cooking time: 20 minutes
Serves: 4 people

- 400g (14oz) fresh egg tagliatelle
- 2 garlic cloves, crushed
- 75g (3oz) pine nuts
- 175g (6oz) fresh or frozen peas
- Handful of fresh mint leaves
- 50g (2oz) Parmesan, grated
- 4 tbsp olive oil
- Salt and pepper, to taste

Cook the pasta in boiling salted water until al dente, then drain.

Place the garlic, one-third of the pine nuts, peas, mint and Parmesan in a food processor and whizz to a paste, gradually adding the oil.

Stir this pesto through the pasta and season.

Sprinkle with shavings of Parmesan, the remaining pine nuts.

LEFT: The anatomy of the humble pea plant is fascinating, with its keeled flowers, delicate yet strong tendrils and pinnate leaves – all of which are edible.

Radish

Raphanus sativus

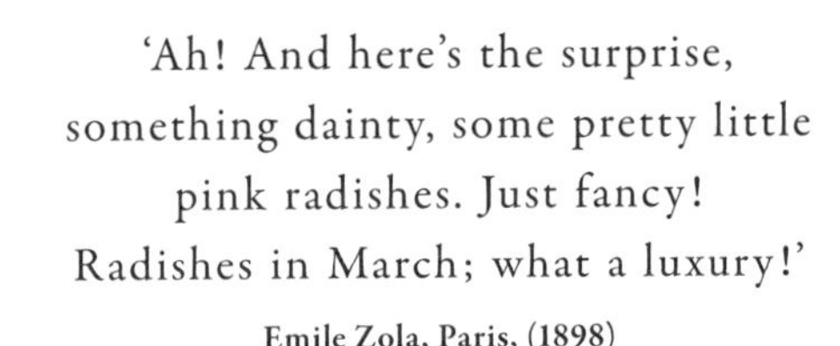

'Ah! And here's the surprise, something dainty, some pretty little pink radishes. Just fancy! Radishes in March; what a luxury!'

Emile Zola, Paris, (1898)

Common name: Radish, common radish, wild radish, garden radish

Type: Annual

Climate: Tender, frost-free winter

Size: 15cm

Origin: Mediterranean

History: There are records of radishes being grown in ancient China. Later they became popular in ancient Egypt where it was the staple diet for the labour community that built the pyramids. The radish was domesticated in Europe in pre-Roman times. The Greek name of the genus *Raphanus* means 'quickly appearing' and refers to the rapid germination of these plants. The common name 'radish' is derived from the Latin meaning root.

Cultivation: Seeds should be sown thinly, ideally in drills 1cm deep, between March and mid April, although they can be sown earlier under cloches. Rows should be 15cm apart. Winter radishes should be sown 25cm apart and sown in late summer.

Storage: Radishes should be picked as soon as they are ready to be eaten because they will not store for much longer than a couple of days. Winter radishes should remain in the ground until needed, although they can be lifted and stored in boxes filled with sand and kept in a cool place.

Preparation: Trim off the tops and root ends. Wash, then slice or grate for use in salads, or cut into decorative shapes for garnishes.

LEFT: Radishes come in all sorts of shapes and colours including the traditional bright red but also white, yellow and pink.

If you are new to growing your own crops, and want to walk before you can run, then this is the crop to try. Not only is it probably the easiest, but it is also one of the quickest to grow, meaning you can test both your gardening and gourmet cookery skills within a few weeks of getting started.

Crunchy, peppery with mustardy overtones, radishes are usually eaten raw and are perfect for pepping up the dreariest and limpest of salads with their vibrant, spicy flavours. It goes best with mild ingredients because otherwise clash with other strong flavours. In salads it goes well with celery,

Growing them as a 'catch crop'

Take advantage of the fast-growing habit of radishes by sowing them quickly in amongst slower growing crops such as onions, potatoes and peas. This avoids wasting valuable space in the garden. They can even be used to mark out rows by planting them in the same drill as parsnips, so you do not lose sight of where this slow-to-germinate winter vegetable is, and so accidentally hoe or weed through them.

Tasting notes

Attractive varieties to try

Brightly coloured radishes to brighten up a salad.

'Sparkler'	A pink variety with a white base.
'French Breakfast 3'	Red with a white tip.
'Ping Pong'	Attractive pure white roots.

Sow these seeds regularly through the season. They only take about 25 days from sowing through to harvesting. Sow little and often about once a week to ensure that there are crops to harvest regularly, rather than sowing one long crop and having to deal with a glut when they are all ready for picking at the same time. If they are left in the ground they quickly turn woody and become unpalatable.

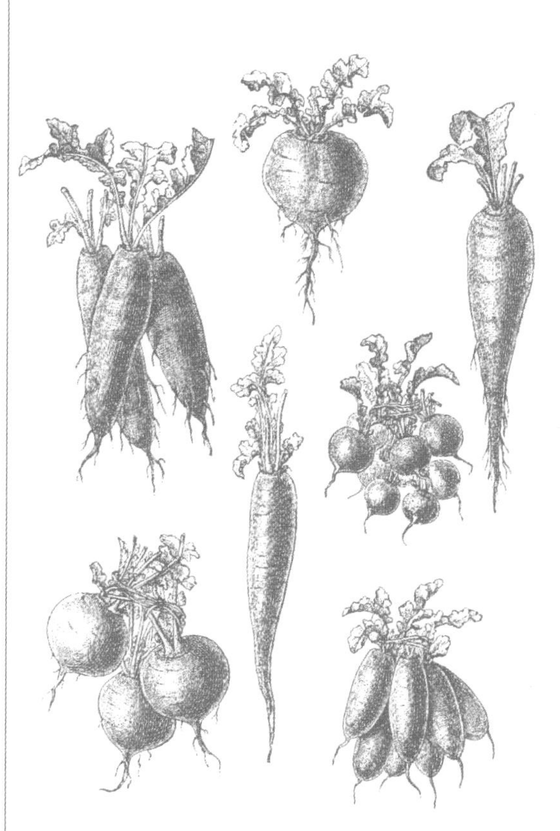

ABOVE: Radishes are one of the quickest growing vegetables in the gardening world. They are also one of the easiest and are simply sown in shallow drills every few weeks during spring and summer.

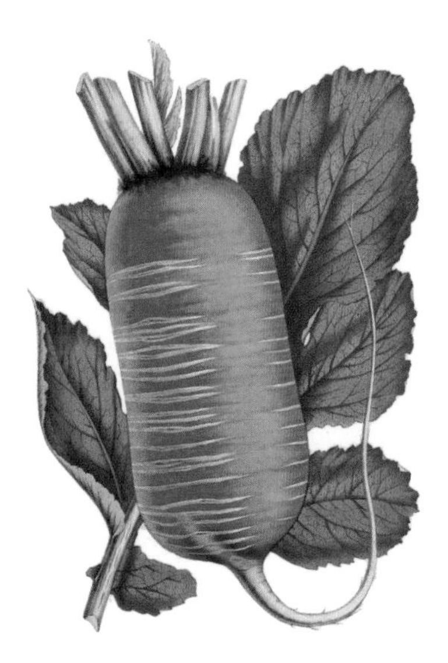

ABOVE: Radishes are the perfect vegetables for the weight watcher as they contain less than 5 calories per serving, meaning you probably expend more calories eating them than are consumed.

beans, crispy lettuce, apples and raisins. It also brightens up poultry dishes such as chicken and turkey. Try a wrap with pea guacamole, turkey and sliced radishes for a simple lunchtime snack that brings out the best of this spicy vegetable's qualities. Interestingly, the heat of radishes is affected by the weather. In hot summers they will taste really spicy, whereas in milder seasons they taste less so. For the weight watchers among aspiring gourmet cooks, radishes are a godsend as they only contain 5 calories per serving. Water is about the only thing less, but the spicy flavours are bound to get that metabolism moving too, as well as burning calories while crunching through it.

It is a root crop and related to mustard, which explains the impressive heat it can mustered from such a small vegetable.

The red-skinned types are the most commonly grown ones in the garden although they come in other colours too. There are also winter radishes, which are hardier and remain in the ground during autumn and winter until needed for harvesting.

Radishes should be grown in a sunny location and only require a shallow soil and one that is not too rich. The soil should also be free-draining but have some moisture-retentive qualities because otherwise they quickly will run to seed and taste woody.

Tasting Notes

Simple radish salad

This is a salad for those who like a hot and spicy side dish. If you want something less intense, then add milder flavoured ingredients from the kitchen garden such as cucumber, tomato or avocado.

Preparation time: 20 minutes
Serves: 2 people (as a side dish)

- 10–12 radishes, thinly sliced
- 1 tbsp salt
- 1/4 tsp ground black pepper
- 2 tbsp fresh lime juice
- 1 tbsp fresh orange juice

Combine the radishes with the salt in a bowl and cover with water. Let it sit for 15 minutes.

Meanwhile, stir together the pepper and fruit juices.

Drain the radishes and then toss in with the dressing.

Add more salt, pepper or fruit juices to taste as needed, and serve.

Mooli

Raphanus sativus var. *longipinnatus*

Common name: Mooli, daikon

Type: Annual

Climate: Hardy, average to cold winter

Size: 15cm

Origin: Asia

History: Closely related to the common summer radish, there are records of radishes being used in ancient China around 800 BC and later in Japan where the long, white daikon radish or mooli became a major food.

Cultivation: Grow them in nutrient-poor soil in full sun. Sow from late summer and harvest in autumn and winter.

Storage: Mooli keep for much longer than summer radishes, and as they are winter hardy they can be kept in the ground until they are needed (usually no later than early winter) in the kitchen. If it is likely that the soil is going to freeze, they can be lifted and stored in boxes of sand or compost.

Preparation: In Japan, mooli is known as daikon and is frequently pickled and served as a crunchy accompaniment to rice at mealtimes. It can also be chopped and put into salads or made into crudités to serve with dips. It can also be steamed and grated or added to stir-fries.

LEFT: Mooli radishes are grown in the same way as winter types and are not suitable for containers.

Good things come to those who wait and this gourmet vegetable is worth waiting for. If you want something to spice up your dishes towards the end of the season then give this giant Asian root vegetable a try.

Mooli is the Hindi word for this large crunchy white radish that is used all over South and East Asia and known as daikon in Japan and by various names in China. They are not as pungent as the smaller radishes and have a high water content. The seed pods are called mongray, and these are eaten as a vegetable too. Mooli is known as 'white ginseng' in China, so is known for its invigorating properties. It purifies the blood and can detoxify the organs and is great for hangovers. In some rural areas of Pakistan it is cultivated for both food and medicine along with other plants such as aloe vera, ajwain, okra, fennel and nightshade (for ear infections).

Mooli is very closely related to the much smaller and more commonly grown red radish that is so popular through the summer months in salads. However, this has a different season of interest, being ready for harvesting as the first autumn frosts arrive in the garden. It gives the gourmet gardener an exciting ingredient to harvest from the vegetable patch that is not one of the stalwart winter crops such as leeks, kale and parsnips. Mooli is far larger than the traditional radish and can be used in a wide range of cuisine. It can be eaten raw but also cooked and is perfect for adding spicy flavours to autumnal stir-fries. It can be sliced raw and used as a radish substitute in salad – alternatively, the root can be grated and used as an ingredient in rostis. It is one of the key ingredients in Kimchi, the Korean fermented pickle, but is also often added to Indian curries and Chinese dishes and is a key ingredient for Dim Sum. The most commonly grown colour of mooli is white but there are lots of other colours to try too.

Like the usual summer radishes, mooli do not like rich soil, too much fertilizer or to have been recently manured, as this causes an excess of leaves to sprout and roots to become distorted and forked. However, due to the larger size of the root they need a much deeper soil than standard radishes, which will happily grow in just a few centimetres of top soil. Before planting dig over the soil thoroughly, breaking up any hard pans just below the surface. Add some grit or sand to ensure there will be a decent root run for them. Rake the soil level and then sow the seeds at about 20cm apart between each plant and between each row. Seeds should be sown in later summer because if they are sown too early they will bolt and quickly turn to seed. After sowing keep the seedlings free from weeds and keep them well watered.

LEFT: *The Daikon and the Baby* by Helen Hyde (1903) depicts radishes being used in ancient China around 800 BC where long white daikon is a major food crop.

Tasting notes

Chinese pickled mooli

Mooli is not has hot as the traditional radish, but its natural crunch makes it a great ingredient for pickling and using as a relish.

Preparation time: 5 minutes, 30 minutes soaking, 8 hours pickling
Serves: 2 people

- 175g (6oz) chopped mooli
- $\frac{3}{4}$ tsp salt
- 1 tbsp rice vinegar
- $\frac{1}{4}$ tsp freshly ground black pepper

In a mixing bowl, toss the mooli with salt. Cover and refrigerate for about 30 minutes.

Drain and rinse, to remove the salt.

Pat dry with kitchen roll, and return to bowl.

Stir in rice vinegar and pepper. Cover and refrigerate for at least 8 hours.

Rhubarb

Rheum × hybridum

Common name: Rhubarb

Type: Perennial

Climate: Hardy, average winter

Size: 1m

Origin: China

History: Rhubarb was first cultivated in Siberia around 2700 BC. It was the dried root of Chinese rhubarb that was highly prized for its medicinal qualities. It was not until the 13th century that Marco Polo brought the root to Europe but little is known of it in Britain until the 14th century. At this time, the price of rhubarb root commanded even more than opium.

Cultivation: Rhubarb can be grown from seed but is easier from crowns or from established plants bought from the garden centre. It needs a rich, fertile soil and should not be harvested the first year after planting. Keep the plant well watered and mulch around its base each year, being careful not to cover up the crown as this can cause it to rot.

Storage: Stems will last for a few days in the fridge but it is best to keep picking them during the season as and when they are required. It can be cooked and then frozen to use in dishes later in the year.

Preparation: Rhubarb is always cooked for eating and can be used in pies, crumbles, fools, puddings and jams. The leaves must not be eaten as they are poisonous. Cut off the leaves, then wash in cold water and chop the stems into cubes. Just eat the red or white parts of the stems – the greener parts of the stem are much tarter.

Nutrition

The stems of the plant contain multiple vitamins and minerals. Rhubarb is a non-dairy source of calcium and promotes healthy bones and teeth. Vitamins A,C, E and K are also present in high levels, helping the body repair and protect its immune system and develop and repair tissues.

BELOW: The emerging shoots of forced rhubarb are a real treat in early spring. The crowns are covered over during winter to encourage them into growth early on in the year.

With just the perfect amount of acidity and sharpness to cut through the sweetness of crumble and custard, yet bursting with flavour, rhubarb is a springtime gourmet treat. One of the big boys of the vegetable garden with its large ornamental leaves and its spreading habit, rhubarb is popular with gourmet gardeners for the beautiful, pink, succulent stems that are forced over winter. Even when the stems have not been forced they are still delicious, with enough sharpness to contrast well with sweet creamy dishes that use yoghurts, ice cream or crème fraiche. Stems must be cooked before eating and rhubarb leaves are poisonous and should not be eaten.

It needs to be grown in rich, fertile soil with lashings of organic matter dug in to feed its huge luxuriant leaves and its rampant growth habit.

TASTING NOTES

Three of the best

If you want rhubarb with the finest taste, then sometimes some of the older, traditional varieties are the best.

'Timperley Early'	The most popular variety for early forcing.
'Champagne'	An old favourite which produces long pink-tinged stems when forced.
'Victoria'	A traditional variety with deep, fleshy stems and superb flavour.

It needs plenty of moisture but it dislikes waterlogged conditions, where the crown will quickly rot. Think carefully about the positioning of the plant – the plants do not mind a bit of shade, and it makes them useful for a shady spot in the garden. Because it is a perennial it will occupy the same spot for a few years, so make sure it is not going to interfere later with crop rotation. Rhubarb is best started from dormant crowns, which should be planted out between autumn and spring. Space plants 75–90cm apart. For those with a small garden or patio, they can be grown in large containers but they should be at least 50cm deep and wide.

To harvest, the stems should be grabbed near the base, twisted and pulled upwards. Harvesting unforced rhubarb usually starts in spring, with the last crops being collected in mid July. Picking any later can harm the plant as it needs to recover for the following year.

After a few years the centre of the crown will need dividing into sections because it becomes woody and congested. To do this, it should be dug out of the ground in autumn and the rootball should be sliced into sections with a sharp spade. The centre should be discarded, but other sections with growing tips should be replanted in the garden. This helps to reinvigorate the plant and is a great way of getting extra plants for free. Allow it a year to recover.

Forcing rhubarb

The best-tasting stems are forced during winter by covering up the crown with clay forcing pots, or alternatively an upturned dustbin. This excludes the light, which forces young, tender shoots to grow upwards. In the gardening world it is called 'forcing'. The bright red stems taste much sweeter and more flavoursome than when grown in normal conditions. The technique of forcing rhubarb into early growth was allegedly discovered at the Chelsea Physic Garden in 1817 when somebody accidentally covered a dormant crown with soil. A few weeks later, when the soil was uncovered, it revealed these delicious blanched stems that were sweeter, redder and better flavoured than when grown under usual conditions. The technique of forcing rhubarb was born. However, it was not until the 1870s that people started growing it commercially, by taking the crowns inside and growing them in the dark in warm forcing sheds. The area that became synonymous with this technique is between Leeds, Wakefield and Bradford and is known as the Yorkshire rhubarb triangle.

Not just a spring treat

Until recently rhubarb has always been something to enjoy from spring until midsummer. However, thanks to a new variety called 'Livingstone' it is now possible to keep harvesting right through until autumn. Summer dormancy, which causes conventional varieties to stop producing stems in midsummer, has been eliminated in this new strain. Now it is possible to combine rhubarb with autumn fruits such as apple and blackberry to create amazing new seasonal combinations of dessert.

With stems being produced by forced rhubarb through winter, and 'Livingstone' producing stems in autumn, it really does seem that rhubarb is not just a spring treat. It is becoming an ingredient for all seasons.

Tasting notes

Rhubarb crumble

Rhubarb crumble is a splendid treat from early spring to midsummer. Ready in the garden before most soft fruits, rhubarb is known as the first 'dessert' ingredient of the year.

Preparation time: 20 minutes
Cooking time: 40 minutes
Serves: 6 people

- 300g (10oz) plain flour
- 150g (5oz) unsalted butter
- 150g (5oz) Demerara sugar
- 250ml (8fl oz) orange juice
- 60g (2oz) caster sugar
- 500g (1lb) rhubarb, chopped

Pre-heat a conventional oven to 180°C (350°F / gas mark 4 / fan 160°C).

Rub together the flour, butter and sugar until mixture resembles breadcrumbs.

Mix the orange juice with the sugar in a pan and bring to the boil. Add the rhubarb, reduce the heat and simmer until it is softened.

Spoon the rhubarb into a baking dish and cover with 6 tablespoons of poaching liquid.

Add crumble mixture on top and bake for 30–40 minutes.

LEFT: Rhubarb is a large vigorous perennial plant with huge leaves, requiring plenty of space in the kitchen garden. It should be mulched each year to retain moisture around the root system.

EXTENDING THE SEASON

It is possible with careful planning to extend the season and to keep the kitchen garden productive for the majority of the year. There are various techniques for doing this including regular or successional sowing and using protection such as cloches or glasshouses to keep plants growing during colder weather.

ABOVE: There are different types of cabbage including spring, summer, autumn and winter types. By selecting the right varieties it should be possible to have cabbages all year round.

SUCCESSIONAL SOWING

Some plants such as lettuces, radishes and carrots can be sown regularly throughout most of the year and will keep producing a crop. Other plants such as cabbages have winter, spring, summer and autumn varieties and will crop depending on when they were sown.

Other examples include:

Broad beans – sow in autumn for early spring crop; sow in early spring for late spring crop.

Carrots – sow from March until September for regular harvesting through that period.

Runner beans – do three sowings, two weeks apart, from mid-May to extend the season.

Garlic and onion sets can be sown in autumn for early summer harvesting and throughout the following months until early spring to stagger the harvest time.

ABOVE: Carrots can be enjoyed from early spring through to the end of the season, if they are sown successionally from early spring.

CLOCHES

The word cloche means bell in French and traditionally were shaped as such so that they could fit snugly over individual plants to protect them either from the spring frosts or from the autumnal cold at the end of the season. Bell-shaped cloches are still commonly used today and are usually made from glass or plastic. If using the latter material it is necessary to peg them down to prevent them blowing away.

For long rows of vegetables, tunnel cloches are a popular option. They usually consist of plastic stretched over hoops and placed over rows of vegetables, a bit like a mini polytunnel. Some cloche tunnels are made from corrugated plastic, which is stronger and less likely to get damaged in the wind.

In hot weather it is important to vent the cloches as otherwise the plants will quickly dry out and the foliage will get scorched in the sun. This is not the case for tunnel cloches with open ends. Cloches will need additional watering compared to plants grown directly outdoors, though plants under plastic or glass will rely on you entirely for their water supply. Cloches are also useful for placing over the soil, prior to sowing, to warm the ground up, which will promote earlier germination in the season.

Make-do cloches

Recycled homemade mini-cloches made from plastic bottles cut in half are a popular option on allotments and are used to protect vulnerable seedlings such as courgettes and pumpkins when first planted out.

COLD FRAMES

These are essentially a wooden or brick box with a sloping glass roof. The roof is usually hinged, enabling plants to be lifted in and out, and meaning it can be easily watered and ventilated on hot days. They are usually used for hardening off plants that have been grown indoors, enabling them to acclimatize for a few days before being planted directly outdoors. This is done by gradually opening the cold frame for progressively longer periods over the course of two weeks. They can also be used like a cloche to extend the growing season.

ABOVE: A wooden-sided cold frame with a sloping glass roof is ideal for hardening off plants that have been started indoors.

GLASSHOUSES

These are a useful addition to any vegetable garden as they enable crops to be sown earlier in the year, getting them off to a good start without having to wait for the weather to warm up outdoors for direct sowing. They also enable more tender crops such as aubergine, chillies and indoor tomatoes to be grown. Glasshouses must have plenty of ventilation as they can quickly get too hot on sunny days. Shade netting can be used to moderate the heat in summer.

Scorzonera

Scorzonera hispanica

Common name: Scorzonera, black salsify, Spanish salsify, viper's grass

Type: Biennial

Climate: Hardy, very cold winter

Size: 35cm

Origin: Mediterranean

History: Scorzonera was first cultivated in the 16th century in Italy and France. The name scorzonera is derived from the Italian words *scorza*, meaning 'bark', and *nera*, meaning 'black'. Also the word scorzone in Italian means a poisonous snake, and the root has been used for a long time to treat venomous snakebites. By 1660, the plant was being cultivated as a vegetable in Italy and France.

Cultivation: This Mediterranean plant is very hardy, but does require a sunny site in well-drained soil. Seeds should be sown in spring for a harvest the following autumn and winter, but they can also be sown in late summer for harvesting the following year. Sow seeds thinly in a shallow drill. Rows should be 30cm apart, and be thinned to 15cm between each plant after germination.

Storage: Keep the roots in the ground until needed for cooking. Once harvested they will store in a fridge for a couple of weeks.

Preparation: Scorzonera is often roasted or boiled, used in a gratin, or in soups. It can be cut into large pieces and cooked, unpeeled. Once cooked, the outer black skin slides off pretty easily. You might want to wear gloves when preparing and cooking scorzonera as it can discolour your fingers.

BELOW: This ancient root vegetable is rarely seen in kitchen gardens these days, but is easy to grow, has attractive flowers, and can be used in a variety of different recipes.

Scorzonera and salsify are often grown together as they are very similar root crops requiring the same growing conditions, both coming from the daisy or *Asteraceae* family. It is still quite rare to find these vegetables either in the shops or restaurants, but they are enjoying a bit of renaissance among foodie fans looking for something a bit different in the culinary and horticultural world. Scorzonera is often referred to as 'black salsify', due to the colour of the outer layer of the root. When the skin is peeled back it reveals an attractive pure white flesh. It is a biennial and if left in the ground for a second season it will produce flowers, and the seeds can be collected and re-sown the following year. They are very easy to grow with no significant pests and diseases. They just need to be kept well watered during dry periods and regularly weeded around to avoid competition for moisture. Roots should be ready for harvesting in autumn, but are fully hardy so can be kept in the ground throughout winter and harvested the following season. Each plant produces one long tap root about 30cm in length. Like parsnips, their flavour is said to be enhanced and sweetened once it has been hit by frosts. Take care not to snap the brittle root when digging it up with a fork. In addition to the root, the plump flower bud can also be picked, steamed and eaten. Flower petals can also be added to salads.

LEFT: Bundles of scorzonera roots make for an unusual harvest. As the roots are quite fragile, it helps to bundle them up to prevent damage.

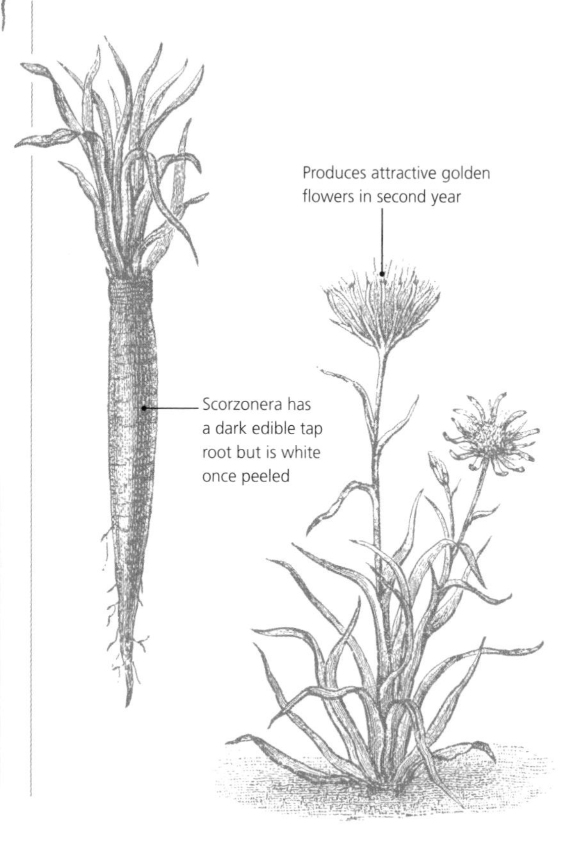

Nutrition

Scorzonera is very low in calories and is a concentrated source of nutrients such as iron, potassium, thiamin (vitamin B1) and phosphorus. It is also a good source of vitamin C, folate, copper and magnesium.

Tomato

Solanum lycopersicum

Common name: Tomato, love apple, pomme d'amour

Type: Annual

Climate: Tender, warm glasshouse

Size: 1.8m

Origin: Peru, Ecuador

History: Tomatoes were first harvested and consumed by the Incas and Aztecs. These tomatoes were the early wild form of the vegetable that grew in the valleys of the Peruvian Andes. It was not until the 16th century that tomatoes were brought over to Europe by Spanish explorers. Tomatoes were not grown in England until the 1590s. John Gerard's *Herbal* published an article in 1597 believing them to be poisonous. The reason it took so long for them to become widely eaten across Britain is because of this misconception of their toxicity. By the mid 18th century tomatoes were popular across England.

Cultivation: Tomatoes can be grown in the glasshouse or outdoors. Seeds are sown in early spring. They should not be planted outside until late spring or early summer, when there is no risk of frost. Plants can be grown in grow bags and will need daily watering during the summer and a liquid tomato feed once a week when the trusses start to form.

BELOW: Tomatoes vary in size and colour from tiny cherry types which can be eaten whole, to large beefsteak types that can be sliced up and either cooked or eaten raw.

Storage: Tomatoes will store for about a week in the fridge but there is no doubt that their flavour is at its best when stored at room temperature on the table, like fruit, and eaten as soon as possible after picking from the vine. They do not freeze well as they end up watery and mushy. Try drying them in the oven and then packing them into jars of olive oil where they can keep for months when stored in a dark, cool cupboard. They can of course be made in tomato sauce or chutney, or made into soup and frozen.

Preparation: If using raw, slice or quarter the tomato. For using in cooked dishes, tomatoes will often need skinning. For this simply cover with boiling water for about 30 seconds, then plunge into cold water and the skins will peel off.

Biting into a sun-ripened tomato picked fresh and warm from the vine is the epitome of Mediterranean living and alfresco dining. There are a whole range of different types to try, from the huge and chunky 'Flame' beefsteaks to the tiny yellow 'Sungold' that simply explode in your mouth with summer sweetness. If ever there was a vegetable that exemplifies the difference in flavour between a shop-bought vegetable (technically a fruit) and one eaten directly from the garden, then this is it. It is used in so many dishes, either uncooked in salads, concentrated, puréed or in sauces. It appears everywhere, from popular Italian favourites such as lasagne, bolognaise and pizza topping through to fish and chip shops with a tub of tomato ketchup. It is used across the world to flavour dishes including Indian, Thai and Mexican. If you want it in a tin then the ubiquitous tomato soup or baked beans must be the two most preserved vegetable dishes in the world.

Just as popular in the garden as the kitchen, tomatoes can be grown in the tiniest of spaces.

Nutrition

Tomatoes are low in sodium, and very low in saturated fat and cholesterol. They are also a good source of vitamins A, C and E and are high in fibre. High levels of the antioxidant lycopene, present in tomatoes, has been shown to lower the risk of cardiovascular disease.

One medium sized-tomato provides over a third of the recommended daily allowance of vitamin C, and nearly a third of the recommended daily allowance of vitamin A.

Tomatoes of love

If you want to cook up a romance with your dinner guest, then make sure your home-grown tomatoes feature strongly on the menu. The French named them *Pomme d'amour* because they believed they were aphrodisiacs. Other vegetables that will supposedly cause more than just a stir in your stir-fry are globe artichoke, lettuce, potato and the turnip, which 'it argumenteth the seed of man and provoketh carnal lust' according to Sir Thomas Elliot in 1539.

They can be grown in growing bags and containers and there are trailing types such as 'Tumbling Tom Red' that are suitable for hanging baskets. In the garden they are a wonderful ornamental addition with the brightly coloured red, yellow and green fruits often making a greater splash than many flowers. For those lucky enough to have a glasshouse, there is a wider range of tomatoes available and the season is extended by a few weeks on either side of summer.

BELOW: Gardeners usually refer to tomatoes as vegetables, yet botanists know of them as a fruit because the seed is contained within their succulent flesh.

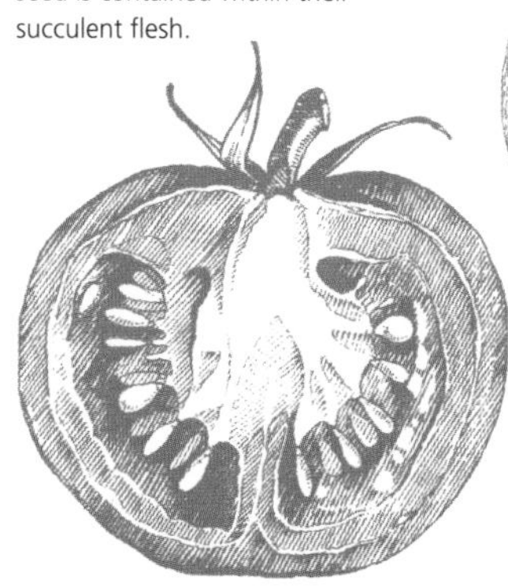

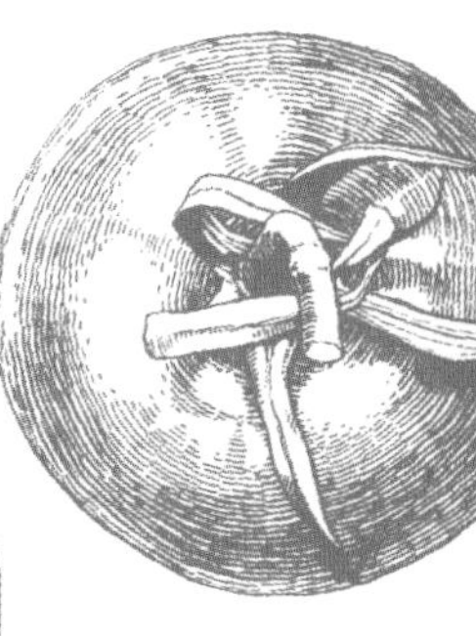

There are still plenty of varieties to choose from to grow outdoors provided that the site is sunny and sheltered and they are kept watered and fed. If planting them directly in the ground then lots of garden compost should be added to the soil. If placing in a growing bag then limit to two tomatoes per bag.

ABOVE: Tomatoes need regular feeding, at least once a month, once they start to produce flowers. Sub laterals should be pinched out to concentrate the plant's energy into the fruit.

TASTING NOTES

Tomato sauce

This tomato sauce should last for a few weeks or even months and provide a supply of sauce to accompany your meat and pasta dishes.

Preparation time: 5 minutes
Cooking time: 20 minutes
Serves: 4 people (for a pasta dish)

- 400g (14oz) chopped skinned tomatoes
- 1 tbsp mixed herbs, chopped
- 1 tbsp tomato purée
- ½ tsp sugar
- 150ml (¼ pint) dry white wine or vegetable stock
- Salt and pepper, to taste
- 1 large garlic clove, crushed

Place all the ingredients in a saucepan.

Bring to the boil, then simmer, uncovered, for 15–20 minutes or until mixture has reduced and thickened. Season.

Tomatoes are usually grown as cordons, with a central leader that is trained up a cane. There are shorter bush types too that do not need a support. Seeds for growing outdoors can be started in March or early April, whereas plants for glasshouses can be started in April. Sow them in small plastic pots indoors and keep them on a sunny window ledge or a propagator. Plant them out in the glasshouse when they are about 20cm tall. Outside varieties should only be planted out after the risk of spring frosts is over. Continue to tie up the central leader to the wire or cane as it grows. When it reaches the top, the growing tip should be pinched out.

Keep the plants watered regularly. Irregular watering will cause the skins to split. Tomato plants should also be given tomato feed once a week. You will soon notice that tomato plants put out lots of sideshoots that extend from the leaf axils. For cordon tomatoes, these must be pinched out as soon as they appear. You do not need to do this with bush types. Water the plants in the evening or morning because otherwise the water will quickly evaporate and splashed leaves could get scorched.

Plants should be picked regularly as soon as they ripen. They will usually turn red although there are different coloured varieties. They feel soft to the touch and when ready, pull away from the vine easily when gently twisted. At the end of the season the green tomatoes can be brought inside and ripened on a sunny ledge. Placing them next to banana skins speeds up the ripening process. If they do not turn red then a green tomato chutney can be made from them.

"TomTato"

If you are short of space in the vegetable garden and cannot decide whether you prefer to grow tomatoes or potatoes, then this could be the answer for you.

A plant called a "TomTato" can be grown which produces tomatoes above the ground, and potatoes below the ground – a potato and tomato on the same plant!

The plants are not genetically modified, they are simply a potato plant and a tomato plant that are grafted together. It works because they are closely related, both belonging to the *Solanaceae* family. The plants last for one growing season, and by the time the tomato harvest is over at the end of summer, the plant can be dug up and the potatoes harvested.

Watch out for the dreaded tomato blight, which causes the leaves and stems to turn yellow and brown before the entire plant rapidly starts to die back. There is very little that can be done to prevent it, although choosing modern, disease-resistant varieties can help. Remove the plants as soon as the symptoms are spotted to prevent the disease from affecting neighbouring plants.

'A world without tomatoes
is like a quartet without violins.'

Laurie Colwin, *Home Cooking*, (1988)

RIGHT: There are so many different types of tomato to try that it is bewildering. Browse specialist seed catalogues and attend summer tastings to find the best ones.

Skirret

Sium sisarum

Common names: Skirret, *suikerwortel* (Netherlands), crummock (Scotland), *zuckewurzel* (Germany)

Type: Perennial

Climate: Hardy, very cold winter

Size: 1m

Origin: Asia

History: Its name comes from the Middle English word skirwhite, perhaps from the Scots *skire*, meaning 'bright', 'clear' and 'white'. Skirret was eaten throughout the Middle Ages and its taste was improved by the addition of wine and honey until it was supplanted by the growing popularity of the potato. By the end of the 18th century, the use of skirrets in cooking had mostly vanished.

Pliny the Elder mentions skirret as being a favourite vegetable of the Emperor Tiberius, who would request a fixed amount of the plant every year from Germany, where it grew especially well despite the cold climate.

Cultivation: This perennial should be planted between autumn and winter. Avoid harvesting its roots in the first year. Plant them at 40cm apart in fertile soil. Their natural habitat is by streams and rivers, so they will need lots of watering during the summer. Harvesting takes place during autumn and winter when the foliage has died down.

Storage: Treat them like carrots. They can be stored in the fridge for a while but can also be placed in 'carrot clamps', which are simply holes dug in the ground outdoors and covered with soil and straw. They can also be kept in boxes or trays covered in slightly moist sand and stored in a garage, shed or cellar.

Preparation: Simply scrub the roots and cut them into suitable lengths for cooking.

LEFT: Skirret is an easy-to-grow perennial root vegetable and is closely related to carrot and parsley. It can be used in many recipes as a carrot substitute,

LEFT: Skirret is a perennial closely related to carrots, and produces an abundance of edible roots underground that can be harvested and cooked.

This one-time popular root vegetable has fallen from grace, and from the dinner table amongst gardeners and cooks alike over the last 200 years. Yet it is remarkably easy to grow and in addition produces attractive plumes of white flowers that look beautiful in the vegetable patch – far prettier than the dull foliage from other root vegetables such as carrots and parsnips. These also play a role in attracting pollinating and beneficial insects such as bees, butterflies and lacewings. The root tastes like a cross between parsnip and potato but is packed full of sweetness. In fact the name 'skirret' is actually derived from the dutch word meaning 'sweet root', *Suikerwortel*.

It is a medium-sized perennial belonging to the same as the carrot and parsley family (*Apiaceae*). It produces sweet and aromatic thin white roots, usually longer than a carrot's and suitable for crunching on raw. In fact, skirret can be substituted for just about any carrot recipe and can be used peeled, sliced and grated in salads. It is also suitable for boiling, roasting, creaming and mashing. Try making a 'carrot' cake from its roots and using it interchangeably with other autumnal vegetables such as parsnips, pumpkins and both sweet potatoes or the usual standard spuds. Like other root vegetables they can also be baked with cooking oil to make chips; although they will be a lot thinner than the chunky chips produced from thicker rooted vegetables such as sweet potato, they will taste just as good.

Skirret prefers full sun, but will grow in dappled shade. It was once a popular waterside plant so it will tolerate a degree of moisture and it prefers a light soil for the roots to be able to penetrate downwards. However, it will benefit from having organic matter added, so add plenty to the planting hole. It is possible to sow seeds of skirret but results can be patchy and it takes a long time to establish. It is far easier to buy young plants from specialist suppliers and herb nurseries for immediate planting. In addition to compost, add sand or grit to the planting hole on heavy soils to help the plant establish well.

Taking root cuttings

If you want more of these plants in the garden they are very simple to divide and propagate from root cuttings. Simply dig up the root system between late autumn and early spring and cut away sections of the plant's roots and replant them else where in the garden. Place the remainder of the plant back in the hole where it should continue to grow.

Avoid harvesting the root in the first year after planting, which will give the plant a chance to send down some roots and get itself established in the garden. The following year, between autumn and winter, once the foliage has died back, dig up the plant and remove sections of the root with a sharp knife before taking it indoors for eating raw or cooking. Do not harvest all the roots or the plant will not survive into spring. Place the plant back in the ground and re-cover the rootball with soil and compost.

Aubergine

Solanum melongena

Common names: Aubergine, egg plant, brinjal, melongene

Type: Annual

Climate: Tender, warm glasshouse

Size: 75cm

Origin: There is debate as to exactly where the aubergine originates from, with some historians believing it comes from India, although it would also appear to have been grown in China in the 5th century.

History: The aubergine was first cultivated over 2,000 years ago in Southeast Asia. In China by 500 BC aubergines had became a culinary favourite to generations of Chinese emperors. The Moors introduced the aubergine to Spain where it received its Catalonian name 'Alberginia'. The vegetable soon spread throughout Europe and by the 16th century Spaniards believed the aubergine to be a powerful aphrodisiac, an 'apple of Love'.

Cultivation: Aubergines require a warm site to survive. Ideally this is in a glasshouse or cold frame but they can survive outdoors in sunny, sheltered spots. Plant two per growing bag and keep them well watered during summer. They can be started by seed in spring, which will give a better choice of variety, but are readily available from garden centres too.

BELOW: Aubergines are easier to grow than most people think. They are usually grown in glasshouses or cold frames unless in a very warm and sheltered garden.

Storage: Aubergines do not last for long and should be used within a few days of harvesting. They do not freeze well because of their high water content.

Preparation: Before cooking, cut off the stems, trim the ends and halve or slice the aubergines. Place in a colander, sprinkling the layers with salt, and leave for 30–45 minutes.

To many people, the aubergine probably looks like more of a curiosity with its funny looking white or purple shaped fruits, rather than something appetizing and appealing to the taste buds. It was treated with scepticism when first introduced to England, not just because of its relationship to the deadly nightshade family (incidentally so are potatoes and tomatoes); it was also blamed for a variety of sicknesses ranging from piles to leprosy. Today it is treated with a similar contempt and sceptical attitude, not for the same reasons but because people think that they are tricky to grow in the garden, and hard to cook in the kitchen. Neither of these is the case, and in the garden it is just as easy to grow as some tomatoes. In the kitchen with a bit of imagination it can be transformed into many sumptuous and simple dishes other than the ubiquitous but delicious moussaka. It can be cooked as part of many other dishes such as ratatouille and as a side dish in its own right. It has a lovely rich texture and can be roasted and puréed. It tastes great when combined with garlic and onions. They can also be thinly sliced and fried to make aubergine crisps. Try stuffing them with a cheese and onion filling or slicing them down the centre and baking them in

ABOVE: The attractive leaves and flowers look very similar to some of its closely related cousins, particularly potatoes and the very poisonous deadly nightshade.

TASTING NOTES

Aubergine varieties to try

There are a huge range of different aubergine types to try. Here are some of the best and most interesting for lovers of gourmet cuisine due to their colours and flavour:

'Pintung Long'	Long lavender-purple fruit packed full of flavour.
'Rossa Bianca'	A gourmet treat with white, rose-tinted fruit with rich creamy flavours and texture.
'Bonica'	Flavoursome fruits with glossy purple skins.
'Galine'	Ornamental and with very tasty, smallish fruit.

the oven with a bed of tomato bolognaise and topped with goat's cheese or mozzarella and herbs.

Aubergines originate in Asia, probably India, and do require warmth and a degree of humidity in the garden. They add a touch of the exotic and beauty to any planting display with their purple-tinted stems and their soft, velvety luxuriant leaves. They also have attractive potato-like bluey flowers but it is their impressive fruit that gives them the wow factor. They are usually dark purple or white, but they also come in other colours including pink, striped, yellow, orange and green. Shapes range from the traditional 'egg shape' to elongated 30cm long types or round. For small patios or balconies, try a dwarf variety in a container. 'Mini Bambino' is a curious form that only reaches 30cm in height and produces tiny 2.5cm fruits.

LEFT: Aubergines are often called egg plants because some varieties have fruits of that shape. They are usually white or dark purple, but can also be stripy, yellow and even pink!

Plants require a warm and sheltered site. Ideally they should be grown in a glasshouse, but they can be grown outdoors in mild areas in full sun. Start seeds off indoors in spring into individual pots or modules. Soaking the seeds overnight before sowing can help with germination. Leave them on a warm window ledge or in a heated propagator. They like warm conditions to germinate, at about 21 to 25°C. Once they have germinated they can be planted into growing bags or containers filled with good-quality compost. If growing outdoors they will need hardening off for about two weeks in a cold frame or porch before moving the growing bags outdoors.

When the plants reach about 40cm high, the growing tips should be pinched out to encourage branching and a bushier habit, as this will promote more fruits. The fruiting sideshoots should be pinched out when no more than five or six fruits have set. The plants will need watering daily during the summer and will benefit from a liquid tomato feed every 10 days from when they start to set fruit. They will need canes and string for support, particularly when the large fruits start to swell.

Aubergines will be ready from midsummer onwards if grown in a glasshouse or cold frame. If they are grown outside then they will not ripen until late summer or early autumn. The fruit should be picked before the skin loses its attractive sheen and glossiness. Use secateurs to cut the fruit from the plant, retaining a bit of the stem on the aubergine.

'That in Italy and other hot countries, where they [the fruits] come to their full maturity, and proper relish, they [the people] doe eate them with more desire and pleasure than we do Cowcumbers.'

John Parkinson, English 17th century horticulturist

Potato

Solanum tuberosum

Common name: Potato, spuds, taters, tatties

Type: Annual

Climate: Tender, frost-free winter

Size: 75cm

Origin: Central America

BELOW: Potatoes are edible tubers produced under the soil. To increase yield and prevent them turning green the soil is earthed up around the stems as the plants start to grow.

History: It is thought that the potato was cultivated in the Andes as early as circa 5000 BC. Spanish adventurers took the domesticated potato back to Europe in the 16th century, although it was not until the late 18th century that it began to become widely consumed.

Cultivation: Potato tubers are planted below the surface of the soil in spring. As they grow, the soil should be 'earthed up' around the base of the shoots to protect them from the frost, to prevent the tubers near the surface turning green, and to increase the potential yield. Potatoes should be grown in a sunny, sheltered site and like a rich soil.

Storage: Early potatoes should be eaten soon after harvesting to capture their unique flavours, although they will keep if stored in paper bags in the dark, such as an under-stairs cupboard. Do not keep them in the fridge or leave them in the light as they will turn green and therefore poisonous. Maincrop potatoes should also be kept in the dark and will store for many months in the right conditions.

Preparation: Wash and scrub the soil off with a brush. Early potatoes and those larger ones intended for baking are usually cooked with their skins on. Maincrop potatoes intended for boiling, mashing and frying are usually peeled. Roast potatoes are usually peeled, but some people think they're crunchier and have better flavour when roasted in their skins.

Ranging from French fries and chunky chips to creamy mashed potato and dauphinois gratin, there is a dish to suit everybody's taste contained within these versatile underground tubers. They are the classic staple food for so many people and can be grown in a variety of different ways. It is hard to imagine the traditional Sunday roast being served without crunchy roast potatoes and equally hard to conceive of a fast-food restaurant without a side order of fries. Whether you crave crisps, waffles or Irish potato bread there is a recipe for everyone. Although they taste delicious and have a wide range of different flavours depending on variety, they are fairly neutral tasting, meaning they make a perfect accompaniment to more powerfully flavoured ingredients, which is why they are the number-one addition to so many dishes. It also explains the popularity of the baked jacket potato filled with just about anything imaginable. Society's dependence on the humble spud reached devastating proportions during the 18th century when many of the Irish starved, partly due to the successive failures of their potato harvest when it became affected by potato blight between 1841–1845.

Nutrition

Potatoes are naturally fat free, a natural source of fibre and contain several vitamins and minerals, such as vitamins C and B6. Vitamin B6 helps to contribute to normal red blood cell formation, normal functioning of the nervous system and the reduction of tiredness and fatigue. All potatoes contain vitamin B6, whatever way they are prepared.

To chit or not to chit

Traditionally, gardeners have always 'chitted' their potatoes to get them off to an early start in the season. This involves starting them into growth about four or five weeks prior to planting by placing the seed tubers in a light, frost-free place to encourage small shoots or 'chits' about 2cm long to form. However, more recently, people are claiming it does not speed up the process, and the warmth of the soil and amount of sunshine are the determining factors in getting an early crop.

Despite its popularity as an edible crop it is related to deadly nightshade, and apart from the underground tubers, the rest of the plant is actually poisonous. However, the tubers themselves are very healthy and contain C and B complex vitamins and are packed full of iron, calcium and potassium. The skins are loaded with fibre and are therefore far more nutritious when cooked unpeeled.

Potatoes are classified into three different groups depending on their season of planting and harvesting, which are first earlies, second earlies and maincrops. First earlies are planted in late winter, second earlies in early spring and maincrops in mid-spring. If space is short, then focus on growing the early varieties as they have the best flavour, take up less area and are expensive to buy in the shops. To plant the tubers, a trench should be dug that is about 15cm deep, with lots of well-rotted organic matter dug into the bottom. Plant the potatoes with their shoots facing upwards (if they have been chitted, see box). Early potatoes should be spaced 30cm apart with 50cm between each row; second earlies and maincrops should be 40cm apart with 75cm between each row. Cover the tubers over gently with a mix of the soil and compost, being

BELOW: Many varieties of potato produce an attractive flower spike. Early varieties are usually ready for harvesting just after the plant has finished flowering.

TASTING NOTES

The perfect roast potato

The perfect roast potato should be crisp and golden on the outside with a fluffy interior. Red-skinned potatoes, such as 'Desirée', are particularly good for roasting, but any floury type like 'Charlotte' will do.

Preparation time: 20 minutes
Cooking time: 50 minutes–1 hour
Serves: 6 people (as a side dish)

- 1kg (2lb) potatoes
- 300g (10oz) goose fat or lard (alternatively sunflower oil or groundnut oil)
- Salt and pepper, to taste

Pre-heat a conventional oven to 240°C (475°F / gas mark 9 / fan 220°C).

Peel and cut the potatoes into egg-sized pieces.

Place in a large pan and pour on cold water until they are just covered and add salt.

Bring to the boil and cook for 5–8 minutes until they are parboiled, then drain.

Toss the potatoes to smash the edges so they become crispy during cooking. Leave to cool.

Place the fat in a large roasting tin and heat in the oven for a few minutes until sizzling hot.

Very carefully tip the potatoes into the tin. Add salt and pepper and turn them over so they are well coated in oil.

Roast in the oven for approximately 45–60 minutes, turning occasionally, until they are soft inside with crispy skins.

Parmentier potatoes

In France, the potato was blamed for causing all sorts of ills, so much so that in 1748 the French government banned its cultivation. It was not until 1772 that the potato was declared fit for human consumption. Convinced that this easily grown tuber could make a useful contribution to the diet of his fellow countrymen, the pharmacist Antoine Parmentier (1731–1813) presented a bouquet of potato flowers to Marie Antoinette, in an attempt to secure royal approval. Today there is a dish called Parmentier potatoes in his honour.

'What can the world do without potatoes? Almost as well might we now ask, what would the world be without inhabitants?'

Samuel Cole, The New England Farmer, (1852)

careful not to break the chits. As the shoots start to emerge from the ground, use a draw hoe or rake to form a ridge in the soil along the row that is between 15 and 20cm high.

There is usually a period of about 12 weeks from planting tubers to harvesting first early types of potato, depending on the quality of the season and the variety. The sign that they are ready for digging up out of the ground is just as they finish flowering, but it is always worth scraping away some of the soil and investigating to see whether the tubers are big enough. Otherwise they can stay in the ground for longer. Maincrop potatoes can be harvested any time after flowering and it is usually at least 20 weeks until they are ready to be lifted from the ground. If they are intended to be stored over winter, they should be left in the ground until the foliage starts to die down. If potato blight strikes, turning the stems black and quickly spreading, then cut down the foliage and begin to harvest the crop. Blight is a problem in wet summers, and if you find it is affecting you year after year, then grow second earlies instead, which should mature before blight strikes.

ABOVE: The potato world has a rich heritage and there have always been lots of varieties to choose from as demonstrated in *Watercolour of Potatoes* by Pierre François Ledoulx c.1790.

ABOVE: On harvesting, you will notice that the original seed potato will have withered, with lots of new potatoes in its place. Take care when harvesting them as they are easily spiked with a fork.

Choosing varieties

There are literally hundreds of potatoes to choose from. Some are more suitable for boiling while others are better for baking or roasting, so choose ones that will suit the type of dishes you enjoy cooking. 'King Edward' and 'Maris Piper' are the traditional best varieties for a baked potato, as well as roasting. 'Red Duke of York' is a versatile variety that can be harvested early and enjoyed like a new potato or left to mature to make a delicious jacket potato with red skin and creamy white flesh. It will come as no surprise to many that the variety 'Golden Wonder' is great for making crisps. It has a dry and floury consistency, but disintegrates when boiled. There are lots of 'salad' type potatoes that have a lovely waxy skin and make perfect salad potatoes. If you suffer from blight in the area, then try a resistant variety such as 'Sarpo Mira'.

Tasting notes

Potato varieties to try

If you think the traditional whitish yellow round potatoes are a bit dull, then there are lots of other colours and shapes for you to use that will brighten up your plate.

'Salad Blue'	There are lots of blue potatoes worth trying. This one is particularly good as it retains its deep bluish colour when roasted and boiled and makes the most amazing-looking purple-blue mash potato.
'Pink Fir Apple'	This quirky looking potato is lumpy and bumpy and has a slightly pinkish tinge to it. It is a maincrop and so can be used well into the Christmas period to accompany the roast turkey. It has a delicious nutty flavour and can be made into the most amazing gourmet chips.
'Highland Burgundy Red'	This red potato retains its red flesh when steamed and roasted. It has a delicious sweet flavour and is one of those curiosities guaranteed to spark a conversation at the dinner table when served up as bright red mash potato.

Spinach
Spinacia oleracea

Common name: Spinach, common spinach

Type: Annual

Climate: Tender, frost-free winter

Size: 25cm

Origin: Ancient Persia

History: The English word 'spinach' dates to the late 14th century, and is from the French word *espinache*, which means 'of uncertain origin'. Spinach is thought to have originated in ancient Persia, and it first appeared in England and France in the 14th century, probably via Spain, and became a popular vegetable. During World War One, wine fortified with spinach juice was given to French soldiers weakened by haemorrhage.

Cultivation: Sow seeds in spring in shallow drills in fertile soil in full sun. Keep the soil moist to prevent the plant from bolting. Sow again in late summer to harvest during autumn and winter. Over-wintering spinach should be protected with horticultural fleece or a plastic cloche to keep the leaves tender.

Storage: Spinach leaves do not last for long at all, and should be picked from the garden or allotment as and when needed that day. They can be frozen for cooking with later in the year.

Preparation: Wash spinach leaves well to remove dirt and pests. Spinach is best steamed for about 5–10 minutes.

RIGHT: Spinach should be sown in early spring but can be prone to bolting and going quickly to seed if the soil is not kept moist during its early stages of development.

Feared by children yet loved by the cartoon character Popeye, spinach alongside Brussels sprouts are the vegetables that everybody loves to hate, and shudder with eye-watering magnitude when placed in front of them on a plate. Yet not only is spinach one of the healthiest vegetables available, packed full of vitamins and antioxidants, it can also taste superb if grown properly in the garden, and prepared well in the kitchen. There are countless dishes that use spinach: a spinach roulade filled with cream cheese and crunchy peppers or spicy chillies is delectable, while spinach and goat's cheese muffins are one of the tastiest savoury cakes out there.

It can be used to make scrumptious Indian dahl soups with yoghurt and served with naan bread or used in a poached egg Florentine breakfast dish. A spinach and ricotta cheese cannelloni is a must for anyone who wants to take their first tentative steps to tasting this gourmet vegetable. Spinach leaves can be eaten raw when young and are often chopped up in salads. But they come into their own when cooked and the leaves break down into a versatile texture that enables them to be spread, baked and easily mixed with other ingredients.

Spinach should be grown in a sunny, sheltered site in well-drained soil. It requires lots of organic matter to help retain moisture, which will help prevent the plant from bolting in dry weather. Sow the seeds thinly in shallow drills 30cm apart, doing so little and often, about every couple of weeks, in short rows to ensure a plentiful supply throughout the season. Once the seeds have germinated they can be thinned out to 15cm between each seedling,

ABOVE: Spinach is a commonly grown leafy vegetable. The young leaves are often eaten raw, whereas the mature leaves are usually boiled, steamed or added to stir-fries.

Nutrition

Spinach is a rich source of omega-3 fatty acids and contain a good amount of minerals like potassium, manganese, magnesium, copper and zinc. Potassium helps the body to control heart rate and blood pressure. Spinach is also a good source of vitamins A, C and K.

100g of farm fresh spinach has 47 per cent of daily recommended levels of vitamin C – a powerful antioxidant, helping the body develop resistance against infectious agents.

'I'm Popeye the sailor man, I'm Popeye the sailor man. I'm strong to the finish because I eats me spinach. I'm Popeye the sailor man!'

'I'm Popeye The Sailor Man', composed by Sammy Lerner, (1933)

TASTING NOTES

Spinach and bacon salad

This simple salad is a mix between the healthy, fresh leaves of young and tender spinach and the crunchy bacon.

Preparation time: 5 minutes
Cooking time: 5 minutes
Serves: 2 people (as a side dish)

- 110g (4oz) spinach leaves, washed
- 170g (6oz) bacon, chopped
- 50g (2oz) croutons
- 1 tbsp cider vinegar

Place the spinach leaves in a bowl.

Fry the bacon in a frying pan until crisp and then add this to the spinach.

Add the vinegar to the pan, stir well to soak up the juices and pour over the salad.

Toss quickly, sprinkle croutons over the top and serve immediately.

but do not throw the seedlings on the compost heap as they can be tossed into salads or used as fillings for sandwiches and wraps.

Sowing can begin in early spring and can continue through to early summer. Avoid sowing during the middle of summer as the heat will cause the plants to bolt. Start sowing again in late summer to provide a crop for overwintering. They will need covering with a cloche when the autumn frosts arrive. This will not just protect the plants but will ensure the leaves are delicious and tender when harvested during those cold winter months.

Harvest the leaves when they are young and tender for salads. They are a quick-growing crop and can usually be picked about 8 weeks after sowing. Do not pick off all the leaves at once, but make regular harvests from it until the plant matures and starts to flower and seed.

RIGHT: Spinach is easy to grow and requires a fertile soil in full sun. Regular, successional sowing will guarantee a plentiful supply of leaves but avoid sowing in summer due to bolting.

Watch out for the fungus-downy mildew, which can cause the leaves to go mouldy in warm, humid conditions. Ensuring the plants are spaced out enough should allow the air to circulate around the foliage, which should hopefully prevent this problem. Alternatively there are modern mildew-resistant varieties.

Also, keep an eye out for birds, which can quickly ruin a crop of spinach seedlings. Either grow seedlings in a fruit cage or cover them with a fine gauge netting or horticultural fleece. Once the plants are established the protection can be removed.

There are a number of varieties to try but 'Atlanta' is one of the hardiest and is ideal for growing through winter. 'Monnopa' is also suitable for autumn or summer, and is useful because it tends not to bolt. It produces thick tasty, strong-flavoured leaves. 'Palco' is another good variety that has some resistance to bolting and also to downy mildew. 'Fiorano' is great for containers and has attractive dark green, rounded leaves. It also has high resistance to downy mildew and good bolting resistanc-making summer sowings possible. 'Bloomsdale Long Standing' is a heritage variety that produces high yields of tender, dark green savoyed leaves.

ABOVE: The hardy annual *Atriplex hortensis* is delicious when added to salads, but also looks interesting in the ornamental garden, with red or green leaves and a purple flower spike.

Alternatives to spinach

Spinach prefers a cool summer and has a tendency to become stressed and bolt, which means it is quick to flower and go to seed. Keeping the plant watered should help to avoid this problem. However, if it continues to be an issue, there are some leafy alternatives. Try Swiss chard or perpetual spinach (see p.48), which has a similar flavour. Alternatively, try growing the more drought-tolerant New Zealand spinach (*Tetragonia tetragonoides*) or mountain spinach (*Atriplex hortensis*), which is simple to grow and the red forms add an attractive splash of colour to the plate.

STORING VEGETABLES

Feast or famine is a common problem for vegetable growers, but by careful planning it should be possible to fill the periods of famine on the plot with vegetables that have been stored during the peak harvest season.

Gluts of fresh vegetables during harvest time are a lovely problem to have on the vegetable plot, but it is not necessary to sling them on the compost heap. One obvious solution is to swap them with friends and family for crops that you may not have grown yourself. Alternatively there are various methods of storing them until ready to use later in the kitchen.

FREEZING

Chest freezers are almost essential for vegetable gardeners these days as there are so many crops that can be frozen. Although most vegetables taste better fresh, some crops such as peas actually taste sweeter when they have been frozen as the freezing process ruptures their cells, imparting more flavour. In addition to freezing vegetables, cooked vegetable dishes can also be placed in the freezer for a delicious instant meal later in the year. Save up plastic boxes and bags so that there are plenty of containers to freeze the vegetables in, and label them clearly.

PRESERVING

Many vegetables can be preserved and made into the most delicious chutneys and pickles (see box). There are fantastic, simple recipes such as picallilli which uses lots of vegetable ingredients from the garden. Onions and beetroot taste delicious when literally stored in a jar of quality vinegar and will keep for years in that form.

LEFT: Most vegetables are best eaten fresh, such as corn on the cob. Sweetcorn also freezes easily, once separated from the cob.

'It is a very good idea to string onions with baler or binder twine. Then hang them in a cool airy place. In many peasant communities the tradition is to hang them against the wall under the eaves of the house.'

John Seymour, *The Complete Book of Self-sufficiency*, (1975)

DRYING

This old traditional method is ideal for preserving chillies and herbs. They can simply be left hanging on strings indoors, with plenty of air circulation, to dry out and be used later in the year in dishes.

Onions and garlic also benefit from being dried out before using. They can be plaited or tied together and left in a dry, rodent-free place for months before using. Alternatively they can be stored in a stocking and hung up on a peg in the garage or shed.

ABOVE: Hardneck garlic does not store for as long as softnecked types. Both types can have their stems plaited together and kept in a dry, frost-free place for a few months.

Sterilizing jars

- First wash the jars and lids in soapy water and rinse in clean, warm water.
- Allow to drip dry, upside-down on a rack in the oven heated at 140°C (275°F / gas mark 1 / fan 120°C) for about 30 minutes.
- Remove by holding with oven gloves.
- Fill with your preserve and cover with lid while still hot.

STORED IN THE DARK OR EVEN THE SOIL

Most vegetables will last longer if stored in a dark cool place, such as a cellar or building, in the shade. Potatoes will keep for longer if kept in a dark cupboard in paper bags.

Some plants can be stored in the ground over winter until they are needed. This is a common method for storing carrots, often called a carrot clamp. A hole is dug in the ground; the carrots are placed in the centre and are then covered over with soil. They are then dug up as and when required. Other root vegetables such as parsnips and salsify can also be stored using this method.

BELOW: Winter squashes, gourds and pumpkins will store for a few months after harvesting if their skin has been cured by leaving to dry in the sunshine for a few days before harvesting.

Chinese artichoke

Stachys affinis

Common name: Chinese artichoke, crosnes, Japanese artichoke

Type: Perennial

Climate: Hardy, average winter

Size: 40cm

Origin: China

History: Chinese artichoke is native to and grown in China and Japan. It is also called the Japanese artichoke. In Europe it is called *crosnes*, which, according to legend, was also the name of a small town in France where the Chinese artichoke was introduced in 1822.

Cultivation: Plant tubers in a fertile, well-drained site in full sun at 25cm apart in early spring. Keep them watered during dry periods and harvest in October. Leave some tubers in the ground to regenerate the following year.

Storage: Keep them in the ground and harvest them as required in the kitchen.

Preparation: Chinese artichoke tubers do not need to be peeled and simply need to be scrubbed prior to eating raw or cooking.

RIGHT: An easy-to-grow plant that is popular in Asia, but very hard to find to find closer to home. They can be added raw to salads or cooked in stir-fries, boiled or steamed.

The tubers of Chinese artichok are a rarity in shops and plant nurseries, which is a shame as these plants are so easy to grow and are the ultimate in gastronomic indulgence. They can be enjoyed raw and crunchy in salads, or stir-fried up in Asian dishes and boiled and steamed. Alternatively their crunchy, nutty texture can be enjoyed when chopped up and fried in garlic butter. They are occasionally found in high-quality or oriental restaurants, but the best way to get your hands

Tasting notes

***Yacon root that tastes a bit like pear* (*Smallanthus sonchifolius*)**

Try yacon for another alternative tuber. It is a large perennial plant, hailing from South America and related to sunflowers, with small yellow flowers. It is simple to grow and requires very little maintenance. In autumn the plant is dug up and the tubers are harvested. Replant some of the tubers with growing tips to ensure there is a crop for the following year.

The tubers have a crunchy texture and their natural sweetness goes nicely when eaten raw in salads (peel tubers first) or added to stir-fries. They make a good substitute for water chestnut recipes. The knobbly growing tips can be divided and replanted, so you do not need to buy additional plants.

The tubers are quite juicy, in fact Yacon means 'water root' in the Inca language. They can be pressed to extract a sweet juice that is then reduced down by boiling to make a sweet syrup, a bit like maple syrup. Scatter nuts over the top and serve immediately.

on these delicious roots is to grow your own. Do not let the look of them put you off; the best way to describe them is like the Australian witchetty grubs with a white, crinkly skin, about 8cm long!

Plant the tubers out between autumn and winter at 8cm deep and 25cm apart in rows. The tubers should be planted horizontally and the rows should be 45cm apart. They are ready for harvesting from October onwards after the foliage has started to blacken and die back. Like Jerusalem artichokes, they can be harvested throughout winter, and once established they should be self-perpetuating so long as some tubers are left in the ground each year after harvesting. Mulch the beds with well-rotted garden compost each spring as this will help to retain the moisture. Water the plants during dry periods.

Oca (*Oxalis tuberosa*)

Tubers are one of the staples of people in Bolivia and Peru. They come in the most amazing bright colours and the taste is like potatoes dipped in lemon juice.

They are very easy to grow as long as you have a moderately long season. Just cook them like you would a potato, either boiled, baked or fried. Tubers start to form in autumn and should be harvested in early winter. The leaves are also edible and delicious.

Salsify
Tragopogon porrifolius

Common name: Salsify, goat's beard, vegetable oyster

Type: Biennial

Climate: Hardy, cold winter

Size: 35cm

Origin: Southern Europe, Mediterranean

History: Salsify was probably first cultivated in Italy in the 16th century but it was not known specifically as a kitchen garden plant. It became very popular in Europe in the late 16th and 17th century and was grown in gardens, together with black salsify (see p.186).

Cultivation: Sow in spring into shallow drills in rows 30cm apart. Thin the seedlings out after germination so that there is 15cm between each plant. They prefer to be grown in full sun in a well-drained soil.

Storage: Salsify is very hardy and the roots can remain in the soil throughout winter until required. Salsify will last two weeks if stored in the fridge, but do not remove their skins.

Preparation: Prepare salsify by scrubbing with a brush, removing the skin, rootlets, and all dark spots. Trim the tops and bottoms and slice as you would a carrot or leave whole. Avoid overcooking this root as it will quickly turn into mush. Salsify can be steamed, sliced and added to soups and stews, or simply mashed and served instead of potato. Salsify goes very well with roasted beef.

RIGHT: Salsify was a popular root vegetable in Europe in the late 16th century, but is less well-known these days. It has a distinctive flavour sometimes compared to oysters.

NUTRITION

Salsify is very low in calories and is a concentrated source of nutrients such as iron, potassium, thiamin (vitamin B1) and phosphorus. Vitamin C, folate, copper and magnesium are also present in high levels, making it an exceptionally healthy vegetable.

Closely related to scorzonera, this root crop is back in fashion in the veggie patch, as well as the kitchen, after years of it being over-looked by cooks and gardeners in favour of carrots and parsnips. Yet, this quirky vegetable really does have a distinctive flavour that explains why it is having a resurgence, and is well worth trying.

The taste is reminiscent of oysters, hence one of their common names 'vegetable oyster', meaning that it goes well with a range of meat and seafood dishes. It also makes a delicious soup by puréeing the roots, adding herbs and spices, and serving it with wholegrain rolls for a satisfying and healthy lunchtime snack.

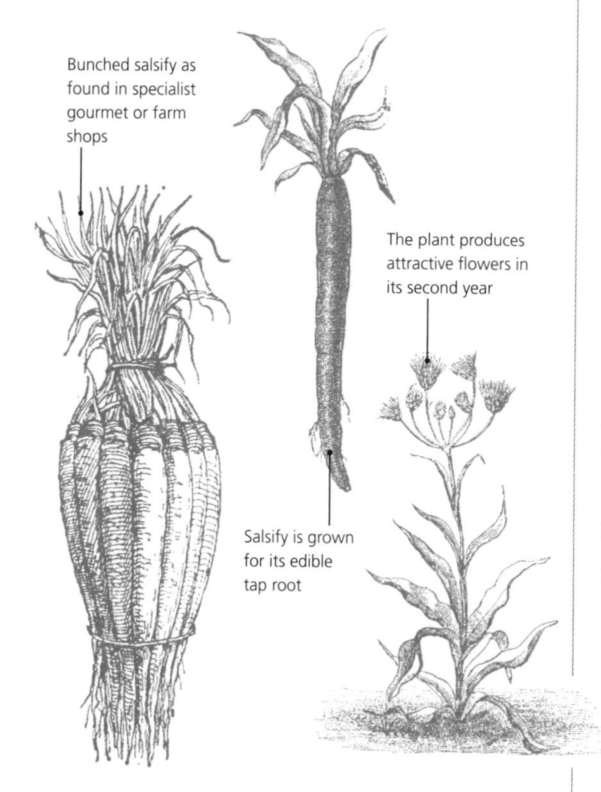

ABOVE: This lesser-known vegetable is a biennial but is usually treated as an annual as the plant is dug up at the end of its first year, in order to harvest the flavoursome roots.

AN ALTERNATIVE ROOT CROP

Hamburg parsley is another unusual root crop worth giving a go in the vegetable patch. It is a hardy plant producing roots that taste like parsnip, but also has leaves that look and taste like parsley. It is the perfect dual-purpose plant.

It makes an ornamental feature in the garden too, as it remains evergreen in winter down to about -10°C, and its attractive foliage is useful for edging paths or growing between rows of vegetables. Sow it in early spring in shallow drills and harvest from late summer.

Salsify is a biennial, meaning it will flower in the second year. However, for most people it is treated as an annual because it is dug up at the end of the first year and the root is harvested. Seeds can be sown from late winter until mid-spring, meaning that the roots will be ready by autumn. They can remain in the ground throughout winter though and if the roots are not large enough they can be left to grow on the following year too. During the growing season, keep the area around the roots watered during dry periods, and weed-free. Harvest as and when required in the kitchen, being careful not to snap the root when digging it out of the ground.

> 'Salsify is a root of high quality, the growing of which is generally considered a test of a gardeners skill.'
>
> **Suttons & Sons, *The culture of Vegetables and Flowers*, (1913)**

Broad bean

Vicia faba

Common name: Broad bean, faba bean, fava bean, horse bean, Windsor bean

Type: Annual

Climate: Half-hardy to hardy, mild to cold winter

Size: 45cm

Origin: Middle East

History: This is one of the most ancient of beans, with a history dating back to the Bronze Age. By the Iron Age broad beans had spread to Europe.

Cultivation: Sow hardy autumn varieties in November for an early harvest the following year. Otherwise sow in late winter or early spring.

Tall varieties may need supports to stop them collapsing. Keep them well watered during dry periods and harvest when the pods feel plump.

Storage: Broad beans are best when shelled and eaten fresh, but they can be frozen. Alternatively, they can be dried and stored in, sterlized sealed jars.

Preparation: When very young and tender the pods can be cooked whole but usually they are shelled. As the plant matures its beans develop a grey outer skin that gradually becomes tougher. This is easily removed after cooking by blanching in boiling water or the beans can be slipped out of their skins after cooking. Cook beans in boiling salted water for about 15–20 minutes.

Nutrition

Broad beans are rich in phyto-nutrients such as isoflavone and plant-sterols. They are also very high in protein and energy as are other beans and lentils. They also contain plenty of health-benefiting antioxidants, vitamins and minerals. Broad beans are also one of the fine sources of minerals like iron, copper, manganese, calcium and magnesium. They also contain high levels of potassium which helps look after the heart and reduce blood pressure.

LEFT: Broad beans develop inside the pod and they can be eaten whole (like a mangetout pea) when very young, but are more commonly shelled and boiled or steamed before eating.

Reckoned to be the oldest and original bean, this is one of the first vegetables of the season is worth savouring and celebrating for that reason alone. You should choose a hardy variety, and good examples are 'Aquadulce' or 'Aquadulce Claudia'. For those with aspirations for self-sufficiency it is an important crop as it fills the gap between the end of the winter crops and the start of the new season. They are usually shelled and can be enjoyed steamed or boiled and are at their best when simply served up with fried onion, garlic and freshly picked mint. They can also be added to soups, stir-fries, casseroles and stews. For a real treat the broad bean pods can be picked when very young and tender, chopped and tailed, steamed and eaten whole. The beans can also be made into a light green purée and when mixed with Dijon mustard make a scrumptious accompaniment to ham, pork or crispy bacon. Alternatively, try mixing with fried garlic, olives and chillies with crumbled feta cheese and served on a crostini.

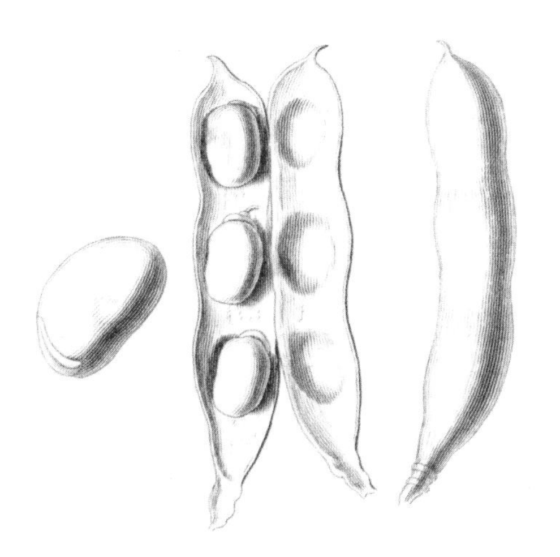

Tasting notes

Broad bean varieties to try

There are lots of different varieties of broad bean to try, but the ones listed below are considered to have the best flavour and are particularly easy to grow.

'Aquadulce Claudia'	This variety is an old favourite among vegetable growers. Good for autumn planting, it produces an early crop of large pods with white beans.
'Medes'	A tall and compact variety with pods producing 5–6 white beans and excellent flavour.
'The Sutton'	A popular dwarf variety that produces an abundance of pods each containing five white tender beans. A good variety for freezing.
'Crimson Flowered'	For a curiosity try this broad bean variety that has deep red flowers followed by small pods filled with small beans of great flavour.

LEFT: Broad beans are one of the oldest beans in cultivation, possibly the original bean. They are also one of the first vegetables to harvest in the new season if sown in autumn.

To get an early crop in spring, seeds can be sown in autumn where they will quickly germinate in the soil. In cooler locations they may need protection with a cloche during winter. Alternatively, they can be sown in late winter or early spring, meaning they will crop later on in the season. To get a regular crop through spring and into early summer, sow seeds once a month. They should be grown in a fertile well-drained site that has been enriched with plenty of organic matter.

Seeds should be sown 5cm deep and 30cm apart in the row, and 45cm apart between rows. They can also be sown in double rows which should be 25cm apart and 60cm between the next row. Rows should be weeded regularly to prevent them competing for nutrients with the seedlings. Growing them in double rows should mean they do not need staking as the plants support each other, but taller varieties in single rows may need tying up with string and stakes being run to the row. The beans will need watering regularly once they start flowering. The tops should be pinched out when the first pods start to form as this will channel their energy into producing beans rather than growing. It also helps mitigate one of their main weaknesses, their susceptibility to black fly, which tend to attack the tip of the plant where it is most tender. For the gourmet gardener the spring tips from these broad beans are worth saving and adding to a stir-fry, but leave them if they have been attacked by black fly, which can be a problem in some years.

The pods should be picked when they start to look plump. Harvest the lowest pods first and do not leave them for too long because the smaller beans are by far the most tender and sweet. Do not forget to pick very young pods to steam and eat whole.

ABOVE: Broad beans are very easy to grow, but require a fertile soil in full sun. Their spring flowers are always a welcome sight as it means your first crop of the year is on its way.

TASTING NOTES

Colour for the plate

Broad beans are usually green or white (although really these are just very light green beans) with the green ones considered to be best for freezing. There are also mahogany red varieties that are worth trying for something different at the dinner table. Try 'Red Epicure' which retains its colour best if steamed rather than boiled.

Sweetcorn

Zea mays

Common name: Sweetcorn, corn on the cob

Type: Annual

Climate: Tender, frost-free winter

Size: 1.5m or more

Origin: Americas

History: Sweetcorn has a very ancient history and is believed to have been domesticated over 8,000 years ago. It was recorded as being the staple food in diets of the Mayan tribes around 2000–1500 BC. In the 16th century it was brought to Europe by the Spaniards.

Cultivation: Sow seeds indoors from mid to late May and plant out after the risk of frosts. They should be planted in a grid system as they are wind-pollinated, so avoid planting in long rows. They require a sheltered, sunny site on fertile soil.

Nutrition

Sweetcorn is gluten free and is low in saturated fat, and very low in cholesterol. Sweetcorn is not particularly nutrient-dense, but does contain folate, which is a B vitamin. Sweetcorn is also surprisingly high in protein and is a good source of fibre, which is essential for bowel health.

Storage: Cobs are best eaten directly after harvesting from the garden, and will not store in the fridge for much more than a few days. The corn can be removed from the cob by scraping it with a sharp knife and stored in bags in the freezer.

Preparation: Remove the outer leaves, silky fibres and stem. If the corn is to be served off the cob then remove it by holding the cob upright and cutting off the corn with a sharp knife, working downwards. Cook in boiled unsalted water for 5–10 minutes or until a kernel comes away from the cob easily.

BELOW: Sweetcorn was introduced to Europe by the Spaniards in the 16th century but its origins can be dated back a lot further with evidence of cultivation over 8,000 years ago.

Fresh cobs of sweetcorn are hard to find in the shops and their season is fleeting. When they do become available they are usually expensive and the taste and the flavour are disappointing. One of the reasons for the blander taste from the shops is because once the cob is cut from the plant, the kernel's natural sugars change to starch and the sweetness is quickly lost. Growing your own plants ensures a plentiful supply and their fresh flavour and sweetness always seem to taste so much better. The simplest way to cook them is simply lightly steamed and then smothered in butter with cracked black pepper, but they can also be used in salads and stir-fries. Yellow varieties are the most common, but it is possible to procure some of the more wacky coloured varieties, some of which are multicoloured. However, they can also be transformed into soups, fritters, chowder and the thick savoury porridge known as polenta. The latter can also be used to make gnocchi.

In the garden they do take up a lot of room, but their sweet flavour definitely makes it worth the effort. They make attractive features with their lofty

Tasting notes

Coloured sweetcorn to try

There are quite a few coloured sweetcorns worth giving a go. They are always a good talking point whether in the garden or around the kitchen table.

'Hopi Blue'	This sweetcorn was developed by the Hopi Indians, a Native American tribe from Arizona. It can be eaten and has a sweet, distinctive flavour. Similar alternatives include 'Blue Jade' and 'Red Strawberry'.
'Bloody Butcher'	This variety has blood-red kernels in wine-red husks. When young it can be eaten like sweetcorn, but as the cob matures it can be ground to make red corn flour.

LEFT: Yellow varieties are the most commonly seen types in the supermarket, but there are a whole range of edible and ornamental corns available including red, blue and multicoloured.

flower spikes and large glossy strap-shaped leaves. Sweetcorn likes a sun-drenched aspect in order for the corns to develop their full sweetness. They need a fertile well-drained soil with lots of garden compost added prior to planting.

Sow sweetcorn at the same time as you are sowing your French and runner beans from mid to late spring. Sow seeds indoors in 9cm pots in general- purpose compost and leave them to germinate on a sunny window ledge or inside the glasshouse. Plant them out after the risk of frost is over, but harden them off first in a cold frame or porch to acclimatize them to the outdoor weather. They should be planted in blocks or grids because they are wind pollinated, meaning that the pollen from the higher male flowers should be able to blow from one plant to another and onto the lower female flowers. Avoid planting them in single rows as they are less likely to get pollinated and therefore will not produce a crop. When the plants are over 1m high the base of the stems can be earthed up. Plants should be kept well watered during dry periods and the area should be kept weed-free, being careful not to damage the shallow roots.

LEFT: To tell whether the corn cob is ready for harvesting, the sheath or outer layers should be peeled back. Push your thumb nail into a kernel, and it will exude a milky sap when ready.

Cobs are ready for harvesting towards the tail end of summer, when the tassel at the end turns brown. Milky sap should squirt out if a fingernail is pushed into the kernel. Cobs should be twisted and pulled to remove them from the plant.

Three sister method

This is a traditional space saving method of growing squashes, beans and sweetcorn in the same space. This method originates from the native North American Iroquois people, who believed these three vegetables were inseparable sisters.

The three sisters technique saves space as three crops are planted together. The squashes sprawl on the ground and their large leaves smother out competing weeds, while the beans use the upright stems of the sweetcorn to train themselves up on. The beans also help prevent the corn from flopping onto the ground, and the roots from the beans fix nitrogen from the air, which helps to sustain the growth of the corn.

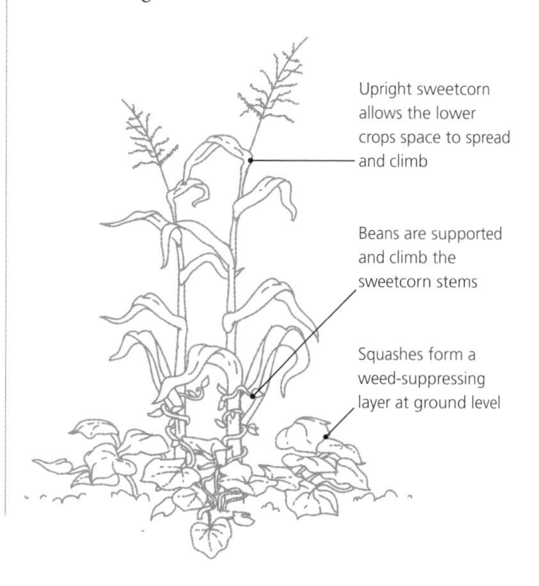

PESTS AND DISEASES

It can be heart-breaking to turn up in the vegetable garden to discover that all your hard work has been destroyed by a pest or infected with a disease. A garden will never be completely free of problems but there are certain precautions that can help reduce them.

GOOD HORTICULTURAL PRACTICE

A healthy plant stands a much better chance of combating pests and diseases than an unhealthy one as its strong growth is more resilient.

- Do not import infected plant material into the garden. Carefully check over plants, whether they are from a garden centre or given by a friend, as a new pest or disease can be very hard to eradicate once it establishes itself.
- Remove diseased plant material as soon as it is spotted. The longer it stays, the more time it has to spread.

CHOOSE RESISTANT VARIETIES

Some varieties have resistance to specific pests or diseases. For example 'Sarpo Mira' potatoes have resistance to blight, and carrot 'Flyaway' has resistance to carrot fly. Careful selection of vegetable varieties could help mitigate many garden problems. Varieties of parsnip such as 'Albion' or 'Palace' have good resistance to parsnip canker.

- Keep on top of the weeds as they will weaken the vegetable plants, depriving them of water and nutrients, which makes them more susceptible to pests and diseases.
- Fungus can spread rapidly if there is poor air circulation between plants, so it is good practice to remove weeds that could be smothering vegetable plants.
- Keep the plants well watered and fed, but avoid overwatering and ensure the foliage does not stay damp as this can cause problems with fungus such as downy mildew.
- Practising crop rotation (see p.154) can reduce a build-up of problems in the soil.

LEFT: Potato blight can be a major problem for vegetable growers, and the fungal disease can rapidly destroy a crop. Foliage should be cut down and removed as soon as blight is spotted to prevent it spreading.

RABBIT PROTECTION

Unfortunately, rabbits like feasting on vegetables just as much as humans and the crops therefore need protecting. Vegetable plots should be surrounded with chicken wire dug down to at least 15cm and turned outwards to prevent rabbits from burrowing underneath.

SLUGS AND SNAILS

Slugs and snails are the number one pest to a vegetable gardener. There are numerous methods of control available, some of which are more effective than others particularly early in the year when there are a lot of young and vulnerable seedlings and juicy new shoots. Slug pellets are usually the most effective control, used sparingly, but feel free to adopt some of these other methods.

- Make a trap, such as an upturned, scooped-out orange or grapefruit, or a half-filled bowl of beer that has been half buried so that the top of it is level with the surface of the soil. The theory is that the slugs and snails fall in the trap and drown.
- Create a barrier: slugs and snails do not like travelling over coarse material such as grit and sand. Surrounding plants with these materials can prevent them from attacking the vegetable plants.
- Nematodes: these are a microscopic organisms, that can be mixed in a watering can to make a solution that is then poured over an infected area. The nematodes attack the slugs and snails, helping to reduce their number.
- Picking them by hand: go out with a torch at night and slugs, snails and caterpillars wlll often by found gorging on plants and seedlings. Dispose of them in salt water.
- Encourage hedgehogs into the garden as they love eating many of the pests. Building a simple hedgehog home in autumn when they are looking for a place to hibernate might be enough to get them to consider making the garden its permanent residence.

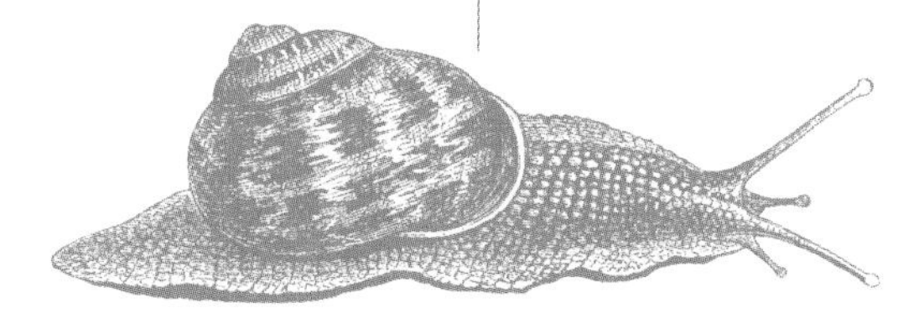

'Shutting one's eyes to trouble due to pests or diseases and hoping for the best is the short cut to serious losses.'

Charles Boff, ***How to grow and produce your own food*****, (1946)**

BIBLIOGRAPHY

Akeroyd, Simon. *A Little Course in Growing Veg and Fruit*, DK, 2013.

Akeroyd, Simon. *The Allotment Handbook*, DK, 2013.

Akeroyd, Simon, Hodge, Geoff, Draycott, Sara, Barter, Guy. *Allotment Handbook*, Mitchell Beazley, Royal Horticultural Society, 2010.

Boff, Charles. *How to Grow and Produce Your Own Food*, Odhams Press Ltd, 1946.

Clevely, Andi. *The Allotment Seasonal Planner*, Collins, 2008.

Davies, Jennifer. *The Victorian Kitchen*, BBC Books, 1989.

Davies, Jennifer. *The Victorian Kitchen Garden*, BBC Books, 1987.

Furner, Brian. *The Kitchen Garden*, Arthur Baker Limited, 1966.

Gammack, Helene. *Kitchen Garden Estate*, National Trust Books, 2012.

Halsall, Lucy. *RHS Step by Step Veg Patch*, DK, 2012.

Harrison, Lorraine. *A Potted History of Vegetables*, Ivy Press, 2011.

Hessayon, Dr.D.G. *The Vegetable and Herb Expert*, Expert Books, 2001.

Klein, Carol. *RHS Grow Your Own Vegetables*, Mitchell Beazley, 2007.

Larkcom, Joy. *Creative Vegetable Gardening*, Mitchell Beazley, 1997.

Laws, Bill. *Spade, Skirret and Parsnip: The Curious History of Vegetables*, Sutton Publishing, 2004.

Pavord, Anna. *Growing Food*, Francis Lincoln Limited, 2011.

Pavord, Anna. *The New Kitchen Garden*, DK, 1996.

Pollock, Michael. *Fruit and Vegetable Garden*, DK, The Royal Horticultural Society, 2002.

Raven, Sarah. T*he Great Vegetable Plot*, BBC Books, 2005.

Shepherd, Allan. *The Organic Garden*, Collins, 2007.

Smit, Tim, Mcmillan-Browse, Philip. *The Heligan Vegetable Bible*, Victor Gollanez, 2000.

Stickland, Sue. *Heritage Vegetables*, Gaia Books Limited, 1998.

Whittingham, Jo. *Grow Something to Eat Every Day*, DK, 2011.

Whittingham, *Jo. Vegetables in a Small Garden*, DK, 2007.

INDEX